W0113967

Elements
of
Tensor
Calculus

A. Lichnerowicz

Translated by
J. W. Leech and D. J. Newman

DOVER PUBLICATIONS
Garden City, New York

Library of Congress Cataloging-in-Publication Data

Names: Lichnerowicz, André, 1915-1998.
Title: Elements of tensor calculus / A. Lichnerowicz ; translated by J.W. Leech and D.J. Newman.
Other titles: Elements de calcul tensoriel. English
Description: Dover edition. | Garden City, New York : Dover Publications, 2016. | An unabridged republication of the 1962 translation of the 4th edition (1958); 1962 translation published by: London : Methuen & Co. Ltd. ; New York : John Wiley & Sons, Inc. | Includes bibliographical references and index.
Identifiers: LCCN 2015047171| ISBN 9780486805177 | ISBN 0486805174
Subjects: LCSH: Calculus of tensors. | Tensor fields. | Geometry, Differential.
Classification: LCC QA433 .L513 2016 | DDC 515/.63--dc23 LC record available at http://lccn.loc.gov/2015047171

Manufactured in the United States of America
80517404
www.doverpublications.com

Contents

PART II: APPLICATIONS

Preface

In 1900 Ricci and Levi-Civita produced a celebrated *mémoire* which gave the first systematic account of tensor calculus and drew the attention of mathematicians and physicists to some of its applications. Since then much has happened. The appearance of the theory of relativity, which would not have been possible without the previous existence of tensor calculus, gave it, in turn, an immense impetus. Tensor calculus has now become one of the essential techniques of modern theoretical physics. It has even been used recently in the study of technical problems such as the interconnection of electrical machines. It can be said that tensor calculus now forms a fundamental part of mathematics and physics.

This little book is divided into two parts, one concerned with tensor algebra and analysis, the other with the most important applications. In Part I the study of tensor algebra ends with a brief consideration of outer product algebras, since this technique deserves to be better known by physicists. On the other hand, the concept of tensor density, which is of little mathematical interest, is not introduced. This concept is, in fact, easily avoided by the introduction of adjoint tensors of antisymmetric tensors.

In the chapters on tensor analysis I have confined myself to the analysis of tensor fields in Riemannian spaces, since Riemannian geometry is the most interesting from the point of view of applications. I have used systematically the method of 'transported reference frames' due to M. Élie Cartan. This method, which is the most geometrical and intuitive, has the added advantage of permitting the reader to avoid the consideration of other generalized geometries.

In the section on applications I was forced to be selective. The first chapter is intended to show the intuitive nature of Riemannian spaces in classical analytical dynamics and their usefulness in this field. In particular, an introduction to the study of continuous media and of elasticity is given. The reader wishing to extend his knowledge in this direction should refer to the excellent works of M. Léon Brillouin.

The remaining two chapters are devoted to the study of Maxwell's equations of the electromagnetic field and to the theory of relativity. Only a brief sketch is given of the principles of general relativity theory. My task will have been accomplished if I have assisted the reader to undertake the study of the fundamental theories of contemporary physics.

Translators' Note

In the course of translation some explanatory footnotes have been added and a number of references to works not available in English have been omitted. The Bibliography has been revised and extended in order to provide suitable suggestions for further reading.

Queen Mary College   J. W. L.
(*University of London*)   D. J. N.

PART I: TENSOR CALCULUS

CHAPTER I

Vector Spaces

I. CONCEPT OF A VECTOR SPACE

1. Definition of a vector space. Consider the set of displacement vectors of elementary vector analysis. These satisfy the following rules:

(*i*) The result of vector addition of any two vectors, $\mathbf{x}$ and $\mathbf{y}$, is their vector sum, or resultant, $\mathbf{x} + \mathbf{y}$. Vector addition has the following properties:

(*a*) $\mathbf{x} + \mathbf{y} = \mathbf{y} + \mathbf{x}$ (commutative property);
(*b*) $\mathbf{x} + (\mathbf{y} + \mathbf{z}) = (\mathbf{x} + \mathbf{y}) + \mathbf{z}$ (associative property);
(*c*) there exists a zero vector denoted by $\mathbf{0}$ such that $\mathbf{x} + \mathbf{0} = \mathbf{x}$;
(*d*) for every vector $\mathbf{x}$ there is a corresponding negative vector $(-\mathbf{x})$, such that $\mathbf{x} + (-\mathbf{x}) = \mathbf{0}$.

(*ii*) The result of multiplying a vector $\mathbf{x}$ by a real scalar α is a vector denoted by $\alpha\mathbf{x}$. Scalar multiplication has the following properties:

(*a'*) $1\mathbf{x} = \mathbf{x}$;
(*b'*) $\alpha(\beta\mathbf{x}) = (\alpha\beta)\mathbf{x}$ (associative property);
(*c'*) $(\alpha + \beta)\mathbf{x} = \alpha\mathbf{x} + \beta\mathbf{x}$ (distributive property for scalar addition);
(*d'*) $\alpha(\mathbf{x} + \mathbf{y}) = \alpha\mathbf{x} + \alpha\mathbf{y}$ (distributive property for vector addition).

Using the above properties as a guide, we now consider a general set E of arbitrary elements $\mathbf{x}$, $\mathbf{y}$ etc., which obey the following rules:

(*1*) *To every pair* $\mathbf{x}$, $\mathbf{y}$, *there corresponds an element* $\mathbf{x} + \mathbf{y}$ *having the properties* (*a*), (*b*), (*c*), (*d*).

(*2*) *To every combination of an element* $\mathbf{x}$ *and a real number* α *there corresponds an element* $\alpha\mathbf{x}$ *having the properties* (*a'*), (*b'*), (*c'*), (*d'*).

We then say that E is a vector space over the field of real numbers and that the elements $\mathbf{x}$, $\mathbf{y}$, *etc., are vectors in E.* If the second rule holds for

1

all *complex* numbers α then E is a vector space over the field of complex numbers. Except when otherwise stated we shall confine ourselves in this book to the study of vector spaces over the field of real numbers.

2. Examples of vector spaces. There are several other simple examples of vector spaces which may be quoted to give an idea of the interest and application of the general concept.

(*a*) Consider the set of complex numbers $a + ib$, where a and b are real. The addition of any two complex numbers ($a + ib$, $c + id$, etc.) and the multiplication of a complex number by a real number α obviously have the properties listed in §1. It follows that the set of complex numbers constitutes a vector space over the field of real numbers.

(*b*) Let **X** be an array of n real numbers arranged in definite order

$$\mathbf{X} = (x_1, x_2, \ldots, x_n)$$

and let E be the set of all arrays **X**. Assume the following two rules of composition:

If $\mathbf{X} = (x_1, x_2, \ldots, x_n)$ and $\mathbf{Y} = (y_1, y_2, \ldots, y_n)$ then

$$\mathbf{X} + \mathbf{Y} = (x_1 + y_1, x_2 + y_2, \ldots, x_n + y_n).$$

If $\mathbf{X} = (x_1, x_2, \ldots, x_n)$ and if α is any real number then

$$\alpha\mathbf{X} = (\alpha x_1, \alpha x_2, \ldots, \alpha x_n).$$

It is easily verified that these two rules imply the rules (1) and (2) of §1. It then follows that E constitutes a vector space with respect to the field of real numbers.

(*c*) Consider the set of real functions of a real variable defined on the interval (0, 1) with the usual composition rules for the sum of two functions and for the product of a function by a constant α. With these rules the set under consideration is a vector space over the field of real numbers.

3. Elementary properties of vector spaces. (*1*) For any two vectors **x** and **y** there is one, and only one, vector **z** such that

$$\mathbf{x} = \mathbf{z} + \mathbf{y}. \tag{3.1}$$

This is easily seen by adding the vector $(-\mathbf{y})$ to each side of (3.1) giving the relation

$$\mathbf{z} = \mathbf{x} + (-\mathbf{y})$$

which defines $\mathbf{z}$ uniquely. As in elementary algebra we write

$$\mathbf{x} + (-\mathbf{y}) = \mathbf{x} - \mathbf{y}.$$

With this notation the property (c') of §1 can be written as

$$(\alpha - \beta)\mathbf{x} = \alpha\mathbf{x} - \beta\mathbf{x}. \tag{3.2}$$

In view of this property it follows that

$$(\alpha - \beta)\mathbf{x} + \beta(\mathbf{x}) = (\alpha - \beta + \beta)\mathbf{x} = (\alpha + 0)\mathbf{x} = \alpha\mathbf{x}.$$

Putting $\alpha = \beta$ it can immediately be deduced from (3.2) that

$$0\mathbf{x} = \mathbf{0}, \tag{3.3}$$

and, on writing $\alpha = 0$

$$(-\beta)\mathbf{x} = -\beta\mathbf{x}.$$

In particular

$$(-1)\mathbf{x} = -\mathbf{x}. \tag{3.4}$$

(2) From (3.4) it follows that the property (d') of §1 can be rewritten in the form

$$\alpha(\mathbf{x} - \mathbf{y}) = \alpha\mathbf{x} - \alpha\mathbf{y}. \tag{3.5}$$

Putting $\mathbf{x} = \mathbf{y}$ in (3.5) it follows that

$$\alpha\mathbf{0} = \mathbf{0}. \tag{3.6}$$

(3) Conversely, the relation

$$\alpha\mathbf{x} = \mathbf{0} \tag{3.7}$$

implies that either $\alpha = 0$ or $\mathbf{x} = \mathbf{0}$. For, if α is not zero it has an inverse α^{-1}, and on multiplying both sides of (3.7) by α^{-1} we have

$$\alpha^{-1}(\alpha\mathbf{x}) = \mathbf{0},$$

or

$$(\alpha^{-1}\alpha)\mathbf{x} = \mathbf{x} = \mathbf{0},$$

which is the required result.

4. Vector sub-spaces. *Definition: A sub-space of a vector space E is any part, V, of E which is such that, if* x *and* y *belong to V and* α *is any real number, then the vectors* x + y *and* αx *also belong to V.*

The commutative, associative and distributive properties of E clearly apply to V. The real number α may be zero so that it is clear that V necessarily contains the zero vector. Again if x belongs to V, $(-1)x = -x$ also belongs to V and it follows that each vector of V has a negative in V. The rules of addition and of multiplication by a scalar thus have the properties listed in §1, therefore V itself is a vector space.

Some simple examples of vector sub-spaces may be given.

(*a*) The set of vectors coplanar with two given vectors constitutes a sub-space of the vector space of elementary geometry.

(*b*) If x is a non-zero vector of a vector space E, the set of products αx, where α is any real number, constitutes a sub-space of E.

(*c*) The set of real functions of a real variable defined on the interval $(0, 1)$ forms a vector space over the field of real numbers. The bounded functions of a real variable defined in the same way form a sub-space of this vector space since the sum of two bounded functions and the product of a bounded function by a constant are themselves bounded.

II. *n*-DIMENSIONAL VECTOR SPACES

5. Basis of a vector space. Let $x_1, x_2, \ldots, x_p$ be p non-zero vectors in a vector space E. These vectors are said to form a *linearly independent* system of order p if it is impossible to find p numbers $\alpha_1, \alpha_2, \ldots, \alpha_p$, not all zero, such that

$$\alpha_1 x_1 + \alpha_2 x_2 + \ldots + \alpha_p x_p = 0.$$

In the contrary case the given system of p vectors is said to be *linearly dependent.*

Consider the set of all systems of linearly independent vectors in the vector space E. There are two possibilities – *either* (*a*) there exist linearly independent systems of arbitrarily large order, *or* (*b*) the order of the linearly independent systems is bounded.

In the second case the vector space is said to have a finite number of dimensions. This classification will be explained shortly. In the remainder of this book we shall only consider *vector spaces which have a finite number of dimensions*. In this case it is possible to determine an integer n such that there exist linearly independent systems of order n but not of order $(n+1)$. If $(e_1, e_2, ..., e_n)$ is any such system of order n it will be called a basis of E in conformity with the following definition:

The basis of a vector space E is any linearly independent system of maximum order.

Let x be any vector in E. The system of $(n+1)$ vectors $(x, e_1, e_2, ..., e_n)$ is necessarily linearly dependent, so there exist $(n+1)$ numbers $\lambda, \alpha_1, \alpha_2, ..., \alpha_n$ such that

$$\lambda x + \alpha_1 e_1 + \alpha_2 e_2 + ... + \alpha_n e_n = 0. \tag{5.1}$$

λ must be different from zero, otherwise the system e_i will not be linearly independent. Equation (5.1) can thus be solved for x and there exist n numbers $x^1, x^2, ..., x^n$ such that

$$x = x^1 e_1 + x^2 e_2 + ... + x^n e_n \tag{5.2}$$

and the vector x is expressible as a linear combination of the e_i. Moreover, this combination is unique for, if there existed another, their difference would constitute a linear combination of the e_i equal to the null vector and with coefficients not all zero. This conflicts with the original postulates. This result may be stated formally:

THEOREM: *Given a basis of E, any vector x of E can be represented in a unique way as a linear combination of the vectors of this basis.*

The numbers $(x^1, x^2, ..., x^n)$ which appear in (5.2) are called the components of x with respect to the basis $(e_1, e_2, ..., e_n)$.

It is easy to show that the property stated in the above theorem specifies bases amongst all systems of vectors. Let $(e_1, e_2, ..., e_p)$ be a system of p vectors such that any vector x of E can be expressed uniquely in the form

$$x = x^1 e_1 + x^2 e_2 + ... + x^p e_p. \tag{5.3}$$

In particular the null vector 0 can be expressed in one way only $(x^1 = x^2 = ... = x^p = 0)$ as a linear combination of the vectors of the system. It follows that such a system must be linearly independent and that $p \leqslant n$.

It is clear that the same property holds for all linearly independent systems of order p. Let $(\epsilon_1, \epsilon_2, \ldots, \epsilon_p)$ be such a system. ϵ_1 can be written

$$\epsilon_1 = \alpha_1 e_1 + \alpha_2 e_2 + \ldots + \alpha_p e_p.$$

Suppose that α_1 (for example) is non zero, then

$$e_1 = \beta_1 \epsilon_1 + \beta_2 e_2 + \ldots + \beta_p e_p,$$

and on substituting in (5.3) it is seen that any vector $\mathbf{x}$ of E can be expressed as a linear combination of $(\epsilon_1, e_2, \ldots, e_p)$. In particular

$$\epsilon_2 = \gamma_1 \epsilon_1 + \gamma_2 e_2 + \ldots + \gamma_p e_p,$$

with at least one of the numbers $\gamma_2, \ldots, \gamma_p$ different from zero, otherwise the system of ϵ_i's cannot be linearly independent. Repeating this procedure we find that any vector $\mathbf{x}$ of E can be expressed in the form

$$\mathbf{x} = \xi^1 \epsilon_1 + \xi^2 \epsilon_2 + \ldots + \xi^p \epsilon_p.$$

It follows immediately that there cannot exist any linearly independent systems of order $(p+1)$, consequently $p = n$. We now express this formally:

THEOREM: *For a system of vectors to constitute a basis of E it is necessary and sufficient that any vector of E can be expressed in one, and only one, way as a linear combination of the vectors of that system.*

The number n is termed the dimension of the vector space under consideration. We shall, in future, use E_n to denote an n-dimensional vector space.

6. Examples. (*a*) In the vector space of elementary geometry a basis is formed by any three non-coplanar vectors. This space is therefore three dimensional.

(*b*) Take the example (*b*) of §2 and consider the vectors

$$\left.\begin{array}{l} e_1 = (1, 0, \ldots, 0), \\ e_2 = (0, 1, \ldots, 0), \\ \cdots \cdots \cdots \cdots \\ e_n = (0, 0, \ldots, 1). \end{array}\right\} \qquad (6.1)$$

Any vector

$$\mathbf{x} = (x^1, x^2, \ldots, x^n)$$

can be expressed in one, and only one, way as a linear combination of the $\mathbf{e}_i$:

$$\mathbf{x} = x^1 \mathbf{e}_1 + x^2 \mathbf{e}_2 + \ldots + x^n \mathbf{e}_n.$$

It follows that these vectors constitute a basis for E, which is thus an n-dimensional space.

7. Vector sub-spaces of E_n. Let V be a vector sub-space of E_n. Then V is a vector space in which any linearly independent system of vectors is also a linearly independent system of E_n. It follows that V has a finite number of dimensions $r \leqslant n$. Let $(\epsilon_1, \epsilon_2, \ldots, \epsilon_r)$ be a basis of V. Any vector of V can be put in one, and only one, way into the form

$$\xi^1 \epsilon_1 + \xi^2 \epsilon_2 + \xi^3 \epsilon_3 + \ldots + \xi^r \epsilon_r. \tag{7.1}$$

Whatever the numbers ξ^1, ξ^2, ..., ξ^r, a vector in the form of (7.1) belongs to V.

Conversely, if $(\epsilon_1, \epsilon_2, \ldots, \epsilon_r)$ is a linearly independent system of order r in E_n it is clear that the set of all vectors which can be expressed as a linear combination of the ϵ_i constitute a vector sub-space of E_n.

Any n-dimensional vector sub-space of E_n coincides with E_n.

8. Complementary vector sub-spaces. Let us consider a linearly independent system $(\epsilon_1, \epsilon_2, \ldots, \epsilon_r)$ of order $r < n$. We wish to complete this system by adding $(n-r)$ new vectors $(\eta_{r+1}, \ldots, \eta_n)$ such that the system of vectors ϵ_i and η_j constitutes a basis of E_n.

There certainly exists in E_n at least one vector such that the system formed by adjoining it to the ϵ_i is a linearly independent system; for if not, the system ϵ_i would itself be a basis of E_n. Let η_{r+1} be this vector so that $(\epsilon_1, \epsilon_2, \ldots, \epsilon_r, \eta_{r+1})$ is a linearly independent system of order $(r+1)$. Repeating this procedure gives systems of increasing order and it will stop only when the order reaches n, the dimensionality of E_n. We now express this formally:

THEOREM: *Given a linearly independent system of order r it is always possible to complete the system with $(n-r)$ vectors to obtain a basis for E_n.*

Let U_r be a vector sub-space of E_n having $r < n$ dimensions and let $(\epsilon_1,\epsilon_2,...,\epsilon_r)$ be any basis of U_r. Using the preceding theorem n vectors $(\eta_{r+1},...,\eta_n)$ can be found (in point of fact in an infinite number of ways) such that the ϵ_i and η_j define a basis of E_n. Let the vector sub-space generated by the η_j be denoted by V_{n-r}.

Evidently the two sub-spaces U_r and V_{n-r} have only the zero vector in common. Moreover any vector $\mathbf{x}$ of E_n which is expressible in the form

$$\mathbf{x} = \sum_{i=1}^{r} \alpha_i\epsilon_i + \sum_{j=r+1}^{n} \beta_j\eta_j$$

can be decomposed into the sum

$$\mathbf{x} = \mathbf{y} + \mathbf{z}$$

of a vector $\mathbf{y}$ of U_r and a vector $\mathbf{z}$ of V_{n-r}. This decomposition is evidently unique according to the preceding remarks. In view of this property the vector sub-spaces U_r and V_{n-r} are said to be complementary. This may be expressed formally:

THEOREM: *To every vector sub-space U_r of E_n there corresponds a unique complementary sub-space V_{n-r}.*

9. Change of basis. It follows from the first theorem of §8 that a vector space E_n has an infinity of bases. We propose to determine the relations that exist between the components of a particular vector $\mathbf{x}$ with respect to two distinct bases.

Let $(e_1,e_2,e_3...,e_n)$ and $(e_{1'},e_{2'},...,e_{n'})$† be two arbitrary bases of E_n. Expressing the vectors of each basis in terms of those of the other we write

$$e_{j'} = \sum_{i=1}^{i=n} A_{j'}^{i}e_i, \qquad e_i = \sum_{j'=1}^{j'=n} A_{i}^{j'}e_{j'}. \tag{9.1}$$

† It is common practice to distinguish bases (and their associated coefficients) by dashes or bars attached to the symbols themselves rather than to the indices. This is simpler to write, but can easily lead to confusion, since the two indices attached to the transformation coefficients refer to different bases. (T.)

Let **x** be an arbitrary vector in E_n having components x^i relative to the basis (e_i) and $x^{j'}$ relative to $(e_{j'})$. We have

$$\mathbf{x} = \sum_{i=1}^{i=n} x^i \mathbf{e}_i = \sum_{j'=1}^{j'=n} x^{j'} \mathbf{e}_{j'} = \sum_{i,j'} x^{j'} A_{j'}^i \mathbf{e}_i. \tag{9.2}$$

Equating the coefficients of the $\mathbf{e}_i$ appearing in the second and last terms of (9.2) gives the transformation formula

$$x^i = \sum_{j'=1}^{j'=n} A_{j'}^i x^{j'}. \tag{9.3}$$

On interchanging the roles played by the two bases we can also write

$$x^{j'} = \sum_{i=1}^{i=n} A_i^{j'} x^i. \tag{9.4}$$

III. DUALITY

10. Linear forms. Let E_n denote a vector space of n dimensions and suppose that to each vector **x** of E_n there corresponds a quantity $F(\mathbf{x})$, the relationship being such that for any **x**, **y** of E_n and for any real α we have

$$F(\mathbf{x}+\mathbf{y}) = F(\mathbf{x})+F(\mathbf{y}), \tag{10.1}$$

$$F(\alpha\mathbf{x}) = \alpha F(\mathbf{x}). \tag{10.2}$$

$F(\mathbf{x})$ *is then said to be a linear form defined on* E_n and (10.1), (10.2) are the relations which characterize such forms.

With the help of (10.1) and (10.2) it is easy to obtain an expression for $F(\mathbf{x})$ in terms of the components of the vector **x** with respect to a basis (e_i). Representing an arbitrary vector of E_n by

$$\mathbf{x} = \sum_{i=1}^{i=n} x^i \mathbf{e}_i$$

it follows, using (10.1) and (10.2) successively, that

$$F(\mathbf{x}) = \sum_{i=1}^{i=n} x^i F(\mathbf{e}_i), \tag{10.3}$$

where the $F(\mathbf{e}_i)$ are independent of $\mathbf{x}$. This may also be written

$$F(\mathbf{x}) = \sum_{i=1}^{i=n} a_i x^i \quad [a_i = F(\mathbf{e}_i)].$$

11. Dual spaces. Consider the set of linear forms defined on E_n and denote its members by y^*, z^*, etc. Assume in addition the following two rules of composition:

(*1*) If $y^*(\mathbf{x}) = \sum y_i^* x^i$ and $z^*(\mathbf{x}) = \sum z_i^* x^i$ denote two arbitrary linear forms, then

$$y^*(\mathbf{x}) + z^*(\mathbf{x}) = \sum_{i=1}^{i=n} (y_i^* + z_i^*) x^i.$$

(*2*) If α denotes any real number, then

$$\alpha y^*(\mathbf{x}) = \sum_{i=1}^{i=n} (\alpha y_i^*) x^i.$$

It is clear that these rules are in accordance with the requirements of §**1**. Hence the set of linear forms defined on E_n constitutes a vector space. Furthermore every linear form can be expressed in one, and only one, way as a linear combination of the n forms $(x^1, x^2, \ldots, x^n)$. This system of n forms thus constitutes a basis for the vector space under consideration which is therefore n-dimensional.

The vector space (E_n^*) *of the linear forms defined on* E_n, *and conforming to* (*1*) *and* (*2*) *above, is called the vector space dual to* E_n.

12. Dual of the basis. Corresponding to each basis $(\mathbf{e}_1, \mathbf{e}_2, \ldots, \mathbf{e}_n)$ of E_n we have set up in a canonical manner a basis $(x^1, x^2, \ldots, x^n)$ of E_n^*. *This will be called the dual of the basis* $(\mathbf{e}_i)$.† If a change of basis is effected on E_n,

$$\mathbf{e}_{j'} = \sum_{i=1}^{i=n} A_{j'}^i \mathbf{e}_i, \qquad \mathbf{e}_i = \sum_{j'=1}^{j'=n} A_i^{j'} \mathbf{e}_{j'}, \tag{12.1}$$

† The dual basis arrived at in the manner indicated obviously depends upon the choice of the vector $\mathbf{x}$. The vector space E_n^* dual to E_n is, however, uniquely defined by E_n. (T.)

a corresponding change of basis defined by (9.3), (9.4) is simultaneously effected on E_n^* such that

$$x^i = \sum_{j'=1}^{j'=n} A_{j'}^i x^{j'}, \qquad x^{j'} = \sum_{i=1}^{i=n} A_i^{j'} x^i. \tag{12.2}$$

Under this change of basis the components y_i^* of the form $y_i^*(\mathbf{x})$ transform according to the relations

$$y_{j'}^* = \sum_{i=1}^{i=n} A_{j'}^i y_i^*, \qquad y_i^* = \sum_{j'=1}^{j'=n} A_i^{j'} y_{j'}^*. \tag{12.3}$$

It is clear that each basis of E_n^* can be considered as the dual of a basis in E_n. Let $(y^{*1'}, y^{*2'}, \ldots, y^{*n'})$ be an arbitrary basis in E_n^*. The forms $(x^1, x^2, \ldots, x^n)$ can be related to this basis as follows

$$x^i = \sum_{j'=1}^{j'=n} A_{j'}^i y^{*j'}.$$

The system of n vectors $\mathbf{e}_{j'}$ defined by

$$\mathbf{e}_{j'} = \sum_{i=1}^{i=n} A_{j'}^i \mathbf{e}_i$$

is linearly independent and consequently forms an alternative basis to $\mathbf{e}_i$ in E_n, for which the dual basis in E_n^* is $(y^{*1'}, y^{*2'}, \ldots, y^{*n'})$.

13. Biduality. Assuming the existence of the vector spaces E_n and E_n^* referred to the dual bases $(\mathbf{e}_i)$ and $(y^{*i} = x^i)$, let us consider the space E_n^{**} dual to E_n^*. Designate by z_i^{**} the elements of the basis in E_n^{**} which is dual to the basis (y^{*i}) in E_n^*. Since the basis (y^{*i}) is dual to $(\mathbf{e}_i)$ and (z_i^{**}) is dual to (y^{*i}), it follows from (12.1) and (12.2) that the $\mathbf{e}_i$ and the z_i^{**} transform according to the same rule. As a consequence, if each vector $\mathbf{x}$ of E_n given by

$$\mathbf{x} = \sum_{i=1}^{i=n} x^i \mathbf{e}_i \tag{13.1}$$

is made to correspond to the element z^{**} of E_n^{**} given by

$$z^{**} = \sum_{i=1}^{i=n} x^i z_i^{**}, \tag{13.2}$$

then this correspondence is *independent of the choice of basis*. It is also manifestly *linear*, in the sense that addition and scalar multiplication remain invariant: to the sum of two elements in E_n there corresponds the sum of two elements in E_n^{**} and to the product of an element in E_n by α there corresponds the product of an element in E_n^{**} by α.

Since there is no intrinsic interest in the elements of E_n^{**} *we find it convenient in what follows to identify the vector space E_n^{**} with E_n and to consider as identical the elements* x *and* z^{**} *which correspond according to (13.1) and (13.2)*. In particular, the basis (z_i^{**}) is identified with the basis $(\mathbf{e}_i)$ so that the dual basis relation becomes a reciprocal one.†

---

† An example drawn from elementary vector analysis may help to clarify the concept of duality.

Let us write

$$F(\mathbf{x}) = \mathbf{F} \cdot \mathbf{x},$$

where **F** is an arbitrary vector. Then the function $F(\mathbf{x})$ clearly satisfies (10.1) and (10.2) and is therefore a linear form defined on 3-dimensional Euclidean space (E_3). Introducing the components of **F** and **x** we have

$$\mathbf{x} = x_1\mathbf{i} + x_2\mathbf{j} + x_3\mathbf{k} \quad \text{and} \quad \mathbf{F} = F_1\mathbf{i} + F_2\mathbf{j} + F_3\mathbf{k},$$

whence

$$F(x) = x_1F_1 + x_2F_2 + x_3F_3.$$

The set of all functions F defines the space, E_3^*, dual to E_3. The elements of E_3^* may alternatively be represented as y^*, z^*, etc., where

$$y^*(\mathbf{x}) = \mathbf{y}^* \cdot \mathbf{x} = x_1 y_1^* + x_2 y_2^* + x_3 y_3^*.$$

These functions y^*, z^*, etc., obviously satisfy (1) and (2) of § 11 and therefore E_3^* is a *vector* space with the basis (x_1, x_2, x_3).

When we go on to define bidual spaces it is clear that the special rule for the formation of elements introduced above cannot be used precisely as before since y^* is not a vector in the normal restricted sense of the term. (It is, of course, an element of E_3^* and therefore a vector according to our general definition.) The elements of the dual space, E_3^{**}, of E_3^* may, however, be defined by

$$z^{**}(y^*) = y_1^* z_1^{**} + y_2^* z_2^{**} + y_3^* z_3^{**}$$

and if we further specify that

$$z_1^{**} = \mathbf{i}, \text{ etc.,}$$

then the basis of E_3 is bidual to itself and E_3^{**} is identical with E_3. (T.)

IV. EUCLIDEAN VECTOR SPACES

14. The summation convention. It is apparent in all the formulae given so far that summations are always effected with respect to an index repeated twice, once as a superscript and once as a subscript. For ease of notation Einstein proposed that in these circumstances the summation sign should be suppressed and the following convention adopted:

The Einstein summation convention: Whenever, in the same term, the same index appears twice, once as a superscript and once as a subscript, a summation is implied over all terms obtained by letting that index assume all its possible values, unless an explicit statement is made to the contrary.

With this convention formulae (9.3) and (9.4) take the form

$$x^i = A^i_{j'} x^{j'}, \qquad x^{j'} = A^{j'}_i x^i, \tag{14.1}$$

whilst formulae (12.3) can be written

$$y^*_{j'} = A^i_{j'} y^*_i, \qquad y^*_i = A^{j'}_i y^*_{j'}. \tag{14.2}$$

The reader will easily convince himself that, far from making algebraic manipulations more difficult to follow, such a convention greatly simplifies the presentation and understanding of the formulae of tensor calculus once some familiarity has been established.

15. Definition of a Euclidean vector space. Consider first the vector space of elementary geometry. For each pair of vectors $\mathbf{x}$, $\mathbf{y}$ the process of scalar multiplication determines a number denoted by $\mathbf{x} \cdot \mathbf{y}$ which is called their scalar product. Scalar multiplication has the following properties:

(a) $\mathbf{x} \cdot \mathbf{y} = \mathbf{y} \cdot \mathbf{x}$ (commutative property);
(b) $(\alpha \mathbf{x}) \cdot \mathbf{y} = \mathbf{x} \cdot (\alpha \mathbf{y}) = \alpha(\mathbf{x} \cdot \mathbf{y})$ (associative property with respect to multiplication by a scalar α);
(c) $\mathbf{x} \cdot (\mathbf{y} + \mathbf{z}) = \mathbf{x} \cdot \mathbf{y} + \mathbf{x} \cdot \mathbf{z}$ (distributive property with respect to vector addition);
(d) If $\mathbf{x} \cdot \mathbf{y} = 0$ for arbitrary $\mathbf{x}$ then $\mathbf{y} = \mathbf{0}$.

Consider, in general, a vector space E_n, defined over the field of real numbers. Suppose there exists a rule of composition which gives for every pair of vectors **x**, **y**, a correspondence with a real number **x**·**y** having the properties (a), (b), (c), (d).

We then say that E_n is a Euclidean vector space and that the rules of composition (a), (b), (c), (d), define scalar multiplication in that space.

It follows from (a), (b), (c), that the scalar product of two vectors is a bilinear form with respect to these vectors; that is a linear form with respect to either of them. From the commutative condition it follows that this bilinear form is symmetric in the two vectors. We now investigate how the scalar product may be expressed analytically.

Let the Euclidean space E_n be referred to an arbitrary basis (e_i) and let

$$\mathbf{x} = x^i e_i, \qquad \mathbf{y} = y^i e_i$$

be any two vectors in E_n. Forming the scalar product and allowing for (b) and (c) we find that

$$\mathbf{x} \cdot \mathbf{y} = x^i y^j e_i \cdot e_j.$$

We are thus led to consider the scalar products of pairs of basis vectors, and find it convenient to write

$$e_i \cdot e_j = g_{ij}. \tag{15.1}$$

Due to the commutative property of the scalar product, the quantities g_{ij} are symmetric in their indices:

$$g_{ij} = g_{ji}.$$

Using this notation we have:

$$\mathbf{x} \cdot \mathbf{y} = g_{ij} x^i y^j.$$

It remains to consider the implications of (d). If $\mathbf{x} \cdot \mathbf{y} = 0$ for any **x** we have for arbitrary values of the x^i

$$g_{ij} x^i y^j = 0.$$

If, in this equation, we put successively $x_1, x_2, \ldots, x_n$ equal to one with the remaining x's zero the following relations are obtained:

$$g_{ij} y^j = 0.$$

These constitute a system of n linear equations in n unknowns which admit only the solutions $y^i = 0$ according to (d). It follows that

$$\det g_{ij} \neq 0,$$

and the bilinear form is said to be non-degenerate. This may be stated as follows:

THEOREM: *The scalar product of two vectors in a Euclidean space is given by the symmetric, non-degenerate, bilinear form*

$$\mathbf{x} \cdot \mathbf{y} = g_{ij} x^i y^j, \tag{15.2}$$

where the g_{ij} represent the scalar products $(\mathbf{e}_i \cdot \mathbf{e}_j)$ of the basis vectors.

It is clear that any sub-space of a Euclidean space is itself Euclidean.

16. Orthogonality and the norm. Let $\mathbf{x}$ and $\mathbf{y}$ be two vectors whose scalar product is zero. In the special case where the Euclidean space under consideration is that of elementary geometry, either $\mathbf{x}$ or $\mathbf{y}$ is zero or they are mutually perpendicular; in either case they are said to be orthogonal. In the general case of an n-dimensional Euclidean space $\mathbf{x}$ and $\mathbf{y}$ are again said to be orthogonal if

$$\mathbf{x} \cdot \mathbf{y} = 0. \tag{16.1}$$

The scalar product of the vector $\mathbf{x}$ by itself is called the *norm* (or *square*) of the vector. We write

$$N\mathbf{x} = (\mathbf{x})^2 = g_{ij} x^i x^j. \tag{16.2}$$

The norm of a vector $\mathbf{x}$ in Euclidean space is thus given in terms of its components by the quadratic form (16.2). A vector whose norm is equal to 1 is said to be *unitary* or *normalized*.

In the space of elementary geometry the norm of a vector is strictly positive and it vanishes only if the vector is zero. If the coefficients g_{ij} of the quadratic form (16.2) can take any value this is no longer true. We are therefore led to the following *definition*:

A vector space is said to be properly Euclidean if it is Euclidean and the norm is strictly positive for every non-zero vector.

In order that the quantity (16.2) be strictly positive for every non-zero vector it is both necessary and sufficient that the quadratic form $g_{ij} x^i x^j$ be positive definite.

17. The Schwarz inequality and its applications. Consider a proper Euclidean space P_n. The norm $N\mathbf{x}$ being always positive or zero, its square root is termed the *modulus* of the vector and we write

$$\sqrt{(N\mathbf{x})} = |\mathbf{x}|.$$

A fundamental inequality exists between the moduli of two vectors of P_n and the modulus of their scalar product:

$$|\mathbf{x}\cdot\mathbf{y}| \leqslant |\mathbf{x}|\,|\mathbf{y}|. \tag{17.1}$$

This is known as the *Schwarz inequality*.

In order to establish this inequality we start with the vector $\lambda\mathbf{x}+\mathbf{y}$, where λ denotes an arbitrary real number. We have

$$(\lambda\mathbf{x}+\mathbf{y})^2 = \lambda^2\mathbf{x}^2+2\lambda\mathbf{x}\cdot\mathbf{y}+\mathbf{y}^2. \tag{17.2}$$

Since the left-hand side is necessarily positive or zero, the discriminant of the quadratic expression in λ appearing on the right-hand side can only be negative or zero. We therefore have the inequality

$$(\mathbf{x}\cdot\mathbf{y})^2 \leqslant \mathbf{x}^2\mathbf{y}^2,$$

which is equivalent to (17.1).

An inequality relating to the modulus of the sum of two vectors can easily be derived from (17.1). Putting $\lambda = 1$ in (17.2)

$$(\mathbf{x}+\mathbf{y})^2 = \mathbf{x}^2+2\mathbf{x}\cdot\mathbf{y}+\mathbf{y}^2.$$

Maximizing the absolute value of $\mathbf{x}\cdot\mathbf{y}$ using the Schwarz inequality gives

$$(\mathbf{x}+\mathbf{y})^2 \leqslant |\mathbf{x}|^2+2|\mathbf{x}|\,|\mathbf{y}|+|\mathbf{y}|^2 = (|\mathbf{x}|+|\mathbf{y}|)^2,$$

hence

$$|\mathbf{x}+\mathbf{y}| \leqslant |\mathbf{x}|+|\mathbf{y}|. \tag{17.3}$$

Finally the Schwarz inequality enables us to define the angle between two vectors of P_n. In elementary geometry the angle ϕ between two vectors is related to their scalar product by

$$\mathbf{x}\cdot\mathbf{y} = |\mathbf{x}|\,|\mathbf{y}|\cos\phi,$$

or

$$\cos\phi = \mathbf{x}\cdot\mathbf{y}/|\mathbf{x}|\,|\mathbf{y}|.$$

Returning to P_n, the Schwarz inequality shows that for two non-zero vectors $\mathbf{x}$, $\mathbf{y}$ of P_n

$$\mathbf{x} \cdot \mathbf{y} / |\mathbf{x}|\,|\mathbf{y}| \leqslant 1.$$

It follows that in the range $(0, \pi)$ there is one, and only one, angle ϕ such that

$$\cos \phi = \frac{\mathbf{x} \cdot \mathbf{y}}{|\mathbf{x}|\,|\mathbf{y}|}. \tag{17.4}$$

This angle is by definition the angle between two vectors $\mathbf{x}$ *and* $\mathbf{y}$ *of* P_n. If $\mathbf{x}$ and $\mathbf{y}$ are defined by their components it follows from (15.2) and (16.2) that

$$\cos \phi = \frac{g_{ij} x^i y^j}{\sqrt{(g_{ij} x^i x^j)}\,\sqrt{(g_{ij} y^i y^j)}}. \tag{17.5}$$

18. Orthonormal systems of vectors. Let P_n be a proper Euclidean n-dimensional space. A system of r vectors of P_n is said to be orthogonal and normalized (or, more briefly, *orthonormal*) if all the vectors in the system are normalized and mutually orthogonal. If $(\mathbf{e}_1, \mathbf{e}_2, \ldots, \mathbf{e}_r)$ are the vectors of the system then

$$\mathbf{e}_i \cdot \mathbf{e}_j = \delta_{ij} \quad (i, j = 1, 2, \ldots, r), \tag{18.1}$$

where $\qquad\qquad \delta_{ij} = 0 \quad \text{if } i \neq j$

$$= 1 \quad \text{if } i = j.$$

It is clear that *every orthonormal system of vectors is necessarily linearly independent.* If this were not true there would exist a relation of the form

$$\sum_{i=1}^{i=r} \alpha_i \mathbf{e}_i = 0, \tag{18.2}$$

where the α_i were not all zero. Suppose α_1, for example, to be different from zero. Taking the scalar product of the members of (18.2) with $\mathbf{e}_1$, we find using (18.1) that

$$\alpha_1 = 0,$$

which is contrary to our hypothesis. It follows that the number of vectors of an orthonormal system is less than or equal to the number of dimensions of the space n. If $r = n$ the orthonormal system under consideration forms an *orthonormal basis* of P_n.

19. The Schmidt orthogonalization procedure. One can ask whether, given an arbitrary integer $r \leqslant n$, orthonormal systems composed of r vectors can be found. We shall now show, using a method due to E. Schmidt, that such systems can indeed be found.

Let $(\mathbf{x}_1, \mathbf{x}_2, \ldots, \mathbf{x}_r)$ be a linearly independent system of order r in P_n and let U_r be the corresponding vector sub-space. It is always possible to construct an orthonormal system whose vectors are linear combinations of the vectors $\mathbf{x}_i$ in U_r. To this end we define the set of vectors $\mathbf{y}_i$ as follows:

$$\left.\begin{aligned}
\mathbf{y}_1 &= \mathbf{x}_1 \\
\mathbf{y}_2 &= \lambda_2^1 \mathbf{y}_1 + \mathbf{x}_2 \\
\mathbf{y}_3 &= \lambda_3^1 \mathbf{y}_1 + \lambda_3^2 \mathbf{y}_2 + \mathbf{x}_3 \\
&\cdots\cdots\cdots\cdots\cdots\cdots \\
\mathbf{y}_r &= \lambda_r^1 \mathbf{y}_1 + \lambda_r^2 \mathbf{y}_2 + \ldots + \lambda_r^{r-1} \mathbf{y}_{r-1} + \mathbf{x}_r
\end{aligned}\right\},$$

where the λ_i^j denote coefficients which we propose to determine in such a way that each $\mathbf{y}_i$ is orthogonal to all the preceding $\mathbf{y}_i$'s. Using the relation

$$\mathbf{y}_2 \cdot \mathbf{y}_1 = 0$$

we deduce

$$\lambda_2^1 \mathbf{y}_1^2 + \mathbf{x}_2 \cdot \mathbf{y}_1 = 0 \quad \text{where } \mathbf{y}_1^2 \neq 0.$$

This determines λ_2^1 and we obtain a $\mathbf{y}_2$ which is both orthogonal to $\mathbf{y}_1$ and non-zero, because the system $(\mathbf{y}_1, \mathbf{x}_2, \ldots, \mathbf{x}_r)$ is linearly independent.

Using the relations

$$\mathbf{y}_3 \cdot \mathbf{y}_1 = 0, \qquad \mathbf{y}_3 \cdot \mathbf{y}_2 = 0,$$

we obtain

$$\lambda_3^1 \mathbf{y}_1^2 + \mathbf{x}_3 \cdot \mathbf{y}_1 = 0 \quad \text{with } \mathbf{y}_1^2 \neq 0;$$

$$\lambda_3^2 \mathbf{y}_2^2 + \mathbf{x}_3 \cdot \mathbf{y}_2 = 0 \quad \text{with } \mathbf{y}_2^2 \neq 0.$$

This determines a vector $\mathbf{y}_3$ which is orthogonal to both $\mathbf{y}_1$ and $\mathbf{y}_2$. It is also non-zero because the system $(\mathbf{y}_1, \mathbf{y}_2, \mathbf{x}_3, \ldots, \mathbf{x}_r)$ is linearly independent according to the definition of $\mathbf{y}_2$.

Continuing in this way a system of vectors $(\mathbf{y}_1, \mathbf{y}_2, \ldots, \mathbf{y}_r)$ is obtained, none of which is zero and which are all mutually orthogonal. Dividing each of these vectors by its modulus we obtain the required orthonormal system of vectors

$$\mathbf{e}_i = \frac{\mathbf{y}_i}{|\mathbf{y}_i|}.$$

20. The space P_n referred to an orthonormal basis. Let a proper Euclidean space P_n be referred to an orthonormal basis $(\mathbf{e}_1, \mathbf{e}_2, \ldots, \mathbf{e}_n)$ so that

$$\mathbf{e}_i \cdot \mathbf{e}_j = \delta_{ij}. \tag{20.1}$$

Writing

$$\mathbf{x} = x^j \mathbf{e}_j$$

and taking the scalar product of each side with $\mathbf{e}_i$, we find

$$x^i = \mathbf{x} \cdot \mathbf{e}_i. \tag{20.2}$$

Since the basis is orthonormal the scalar product of two vectors is given by

$$\mathbf{x} \cdot \mathbf{y} = x^1 y^1 + x^2 y^2 + \ldots + x^n y^n, \tag{20.3}$$

and the norm of a vector becomes

$$N\mathbf{x} = (x^1)^2 + (x^2)^2 + \ldots + (x^n)^2. \tag{20.4}$$

Equations (20.3) and (20.4) will be recognized as the generalization to n-dimensions of the familiar formulae of elementary vector analysis.

We note, moreover, that it is always possible to bring a given positive definite form $g_{ij} x^i x^j$ into the form of (20.4) by a suitable change of basis.

21. Contravariant and covariant components of a vector. Let E_n denote a Euclidean vector space which is referred to an arbitrary basis $(\mathbf{e}_1, \mathbf{e}_2, \ldots, \mathbf{e}_n)$. We have seen in the preceding section that, if this basis is orthonormal, the components x^i of the vector x are equal to the scalar products of x by the vectors of the basis. If the basis is not orthonormal this is no longer true. We are thus led to the following definition.

DEFINITION: *Given an arbitrary basis* $(e_1, e_2, ..., e_n)$ *of the Euclidean vector space* E_n,

(1) *the term* contravariant components *of a vector* x *referred to this basis is applied to the numbers* x^i *which are such that*

$$\mathbf{x} = x^i \mathbf{e}_i, \tag{21.1}$$

(2) *the term* covariant components *of a vector* x *referred to this basis is applied to the scalar products*

$$x_i = \mathbf{x} \cdot \mathbf{e}_i. \tag{21.2}$$

In what follows, contravariant components are always represented by means of superscripts and covariant ones by subscripts. The reasons for this distinction will appear shortly (§23).

The covariant components of a vector may be determined readily from a knowledge of its contravariant components. From (21.1) we have

$$\mathbf{x} = x^j \mathbf{e}_j.$$

Forming the vector product of each side with $\mathbf{e}_i$ it follows that

$$x_i = \mathbf{e}_i \cdot \mathbf{e}_j x^j,$$

which becomes, on introducing the g_{ij}

$$x_i = g_{ij} x^j \quad (i = 1, 2, ..., n). \tag{21.3}$$

Conversely, let us seek to express the contravariant components of a vector in terms of its covariant components. We require the solution of the system of n linear equations in n unknowns x^j given by

$$g_{ij} x^j = x_i. \tag{21.4}$$

From the considerations of §15 the determinant of the g_{ij} is different from zero and the system (21.4) is therefore a Cramer's system.† In

† A system of linear equations with a unique solution. (T.)

the following we write the determinant of the g_{ij} as g and the coefficient of g_{ij} in the development of g as α^{ji}; using Cramer's rule, it follows that

$$x^j = \frac{\alpha^{ji}}{g} x_i.$$

Writing $$g^{ji} = \frac{\alpha^{ji}}{g} \qquad (21.5)$$

we get the fundamental relations

$$x^j = g^{ji} x_i. \qquad (21.6)$$

Since the determinant formed from the elements g_{ij} is symmetric, we have $\alpha^{ji} = \alpha^{ij}$ and consequently the quantities g^{ij} defined by (21.5) are also symmetric with respect to their indices. Moreover, from a well-known result of the theory of determinants

$$\det \alpha^{ji} = g^{n-1},$$

hence $$\det g^{ij} = \frac{1}{g}. \qquad (21.7)$$

22. Expressions for scalar product and norm in terms of covariant components. The use of covariant components leads to particularly simple expressions for the scalar product and norm in terms of the components referred to an arbitrary basis.

Substituting (21.3) into the relation

$$\mathbf{x} \cdot \mathbf{y} = g_{ij} x^i y^j$$

it follows that

$$\mathbf{x} \cdot \mathbf{y} = x_i y^i = x^i y_i. \qquad (22.1)$$

We also find, for the norm of a vector $\mathbf{x}$,

$$N\mathbf{x} = x^i x_i. \qquad (22.2)$$

On the other hand, the scalar product of two vectors and the norm of a vector can be expressed uniquely in terms of the covariant components of the vectors. Using (21.6) we have

$$\mathbf{x} \cdot \mathbf{y} = g^{ij} x_i y_j \qquad (22.3)$$

and $$N\mathbf{x} = g^{ij} x_i x_i \qquad (22.4)$$

Because of (21.7) the bilinear form (22.3) and the quadratic form (22.4) are both non-degenerate.

23. The effect of a change of basis on the contravariant and covariant components of a vector. We have just seen that if the Euclidean vector space is referred to an arbitrary basis the scalar product of two vectors $\mathbf{x}$, $\mathbf{y}$ can be expressed in the form

$$\mathbf{x} \cdot \mathbf{y} = x_i y^i.$$

If $\mathbf{x}$ is a constant vector, the scalar product $\mathbf{x} \cdot \mathbf{y}$ defines a linear form with respect to the arbitrary vector $\mathbf{y}$. This linear form has the components x_i relative to the basis $(y^1, y^2, \ldots, y^n)$ which is dual to $(\mathbf{e}_1, \mathbf{e}_2, \ldots, \mathbf{e}_n)$.

Consider a change in the basis of E_n defined by

$$(a)\ \mathbf{e}_i = A_i^{j'} \mathbf{e}_{j'}, \qquad (b)\ \mathbf{e}_{j'} = A_{j'}^i \mathbf{e}_i. \qquad (23.1)$$

We have seen that the contravariant components x^i of a vector transform according to the formulae

$$(a)\ x^i = A_{j'}^i x^{j'}, \qquad (b)\ x^{j'} = A_i^{j'} x^i. \qquad (23.2)$$

As for the covariant components x_i, we have seen that these transform according to the formulae (12.3) for the components of a linear form, thus

$$(a)\ x_i = A_i^{j'} x_{j'}, \qquad (b)\ x_{j'} = A_{j'}^i x_i. \qquad (23.3)$$

It will be noted that the linear expressions appearing on the right-hand side of (23.1) (a) and (23.3) (a) contain the same coefficients. This is also true of (23.1) (b) and (23.3) (b). On the other hand, (23.2) (a) and (23.1) (b) have a similar form. These similarities justify our use of the terms *co*variant and *contra*variant.

24. Euclidean vector spaces and duality. If $\mathbf{x}$ is a vector in the Euclidean vector space E_n we have seen that the scalar product $\mathbf{x} \cdot \mathbf{y}$, where $\mathbf{y}$ is an arbitrary vector, enables a correspondence to be set up between the vector $\mathbf{x}$ and a linear form defined on E_n. Conversely, any linear form defined on E_n can be considered as the scalar product of a fixed vector $\mathbf{x}$ with an arbitrary vector of E_n.

If the space E_n is referred to an arbitrary basis (e_i) and the dual space E_n^* to the dual basis (x^i), the element in E_n which has the components y^i corresponds to the element in E_n^* with the components y_i. These components are linked by the relations (21.3) and (21.6), i.e.

$$y_i = g_{ij}y^j, \qquad y^j = g^{ij}y_i. \tag{24.1}$$

Thus, the existence of the scalar product enables us to set up a one-to-one correspondence between the elements of E_n and E_n^*, and the components of these elements bear a simple relationship to one another. The Euclidean vector space E_n and its dual E_n^* are therefore considered to be equivalent.

CHAPTER II

Affine Euclidean Point Spaces

25. Definition of an affine space. The points of the space $\mathscr{E}$ of elementary geometry define vectors – the position vectors of elementary vector analysis. These obviously satisfy the relations

$$\overrightarrow{AB} = -\overrightarrow{BA},$$

$$\overrightarrow{AB} = \overrightarrow{AC} + \overrightarrow{CB}.$$

Choose an arbitrary point O in $\mathscr{E}$, then each point A of $\mathscr{E}$ is associated with a displacement vector **a** defined by

$$\mathbf{a} = \overrightarrow{OA}.$$

Conversely, given an arbitrary displacement vector **a**, there exists a point A such that

$$\overrightarrow{OA} = \mathbf{a}.$$

In general, consider a manifold of points, $\mathscr{E}$, and suppose that to each pair (A, B) of points of $\mathscr{E}$ taken in order there corresponds a vector of an n-dimensional vector space E_n denoted by $\overrightarrow{AB}$. Assume this correspondence to have the following properties:

(a) $\overrightarrow{AB} = -\overrightarrow{BA}$;
(b) $\overrightarrow{AB} = \overrightarrow{AC} + \overrightarrow{CB}$;
(c) if O is an arbitrary point in $\mathscr{E}$ then, to every vector **a** of E_n, there corresponds a unique point A such that

$$\overrightarrow{OA} = \mathbf{a}.$$

When these conditions hold we say that the manifold $\mathscr{E}$ is an affine point space of n-dimensions. We shall denote it henceforth by the symbol $\mathscr{E}_n$. The space $\mathscr{E}_n$ will be termed real or complex according to whether the associated vector space E_n is defined over the real or complex field. We shall limit future considerations to real affine spaces.

Given a vector denoted by $\overrightarrow{AB}$ we say that A is the origin and B the extremity. Condition (c) consequently states that, for all $\mathbf{a}$, the specification of $\mathbf{a}$ in E_n and an origin O in $\mathscr{E}_n$ completely determines the extremity A.

From the above hypotheses it is clear that

$$\overrightarrow{AA} = \mathbf{0}.$$

26. Reference frames of affine spaces. DEFINITION: *Given an affine space $\mathscr{E}_n$, then an arbitrary point O of $\mathscr{E}_n$ and a basis $(\mathbf{e}_1, \mathbf{e}_2, \ldots, \mathbf{e}_n)$ of the associated vector space together constitute a reference frame of $\mathscr{E}_n$. The point O is called the origin of the reference frame.*

A reference frame will henceforward be represented by the notation $(O, \mathbf{e}_1, \mathbf{e}_2, \ldots, \mathbf{e}_n)$ or, more concisely, by $(O, \mathbf{e}_i)$.

Let A denote an arbitrary point of $\mathscr{E}_n$. The components of the vector $\overrightarrow{OA}$ with respect to the basis $(\mathbf{e}_i)$ are called the *coordinates* of A in the reference frame $(O, \mathbf{e}_i)$. By virtue of (c) there is a unique correspondence between the sets of n real numbers $(x^1, x^2, \ldots, x^n)$ and the points A of $\mathscr{E}_n$.

Given two points A and B of $\mathscr{E}_n$, defined by their coordinates (x^i) and (y^i) in the reference frame $(O, \mathbf{e}_i)$, we wish to define the vector $\overrightarrow{AB}$. From (a) and (b) we have

$$\overrightarrow{AB} = \overrightarrow{AO} + \overrightarrow{OB} = \overrightarrow{OB} - \overrightarrow{OA}.$$

From this we deduce that the vector $\overrightarrow{AB}$ has the n quantities $(y^i - x^i)$ as components with respect to the basis $(\mathbf{e}_i)$.

27. Change of reference frame. We wish to determine the relations that exist between the coordinates of a given point M of $\mathscr{E}_n$ referred to two distinct frames.

Let (O, e_i) and $(O', e_{j'})$ be two reference frames of $\mathscr{E}_n$. In order to locate each frame with respect to the other we refer the origin of each frame to the other and the vectors of each basis to the other basis. We have

$$\overrightarrow{OO'} = \alpha^i e_i; \qquad \overrightarrow{O'O} = \alpha^{j'} e_{j'};$$

$$e_{j'} = A_{j'}^i e_i; \qquad e_i = A_i^{j'} e_{j'}.$$

Let M be an arbitrary point of $\mathscr{E}_n$ whose coordinates are x^i in the frame (O, e_i) and $x^{j'}$ in $(O, e_{j'})$. We have

$$\overrightarrow{OM} = x^i e_i \tag{27.1}$$

also

$$\overrightarrow{OM} = \overrightarrow{OO'} + \overrightarrow{O'M} = \alpha^i e_i + x^{j'} e_{j'} = (\alpha^i + A_{j'}^i x^{j'}) e_i. \tag{27.2}$$

Identifying the right-hand sides of (27.1) and (27.2), and comparing coefficients of e_i, we get the transformation formulae

$$x^i = \alpha^i + A_{j'}^i x^{j'}. \tag{27.3}$$

Interchanging the roles of the two reference frames we can also write

$$x^{j'} = \alpha^{j'} + A_i^{j'} x^i. \tag{27.4}$$

28. Affine sub-spaces. DEFINITION: *Let $\mathscr{V}$ be a part of an affine space $\mathscr{E}_n$ such that for any point O in $\mathscr{V}$ the set of vectors $\overrightarrow{OM}$ (all M in $\mathscr{V}$) constitutes a vector sub-space of E_n. $\mathscr{V}$ is then said to form an affine sub-space of $\mathscr{E}_n$.*

A sufficient condition for $\mathscr{V}$ to be an affine sub-space is that, for a particular point O, the vectors $\overrightarrow{OM}$ constitute a vector sub-space V_r of E_n. If O' is any other point in $\mathscr{V}$

$$\overrightarrow{O'M} = \overrightarrow{OM} - \overrightarrow{OO'}$$

and since $\overrightarrow{OM}$ and $\overrightarrow{OO'}$ both belong to V_r, so must $\overrightarrow{O'M}$. Conversely, given a vector **a** of V_r, there exists a unique point M belonging to $\mathscr{V}$ which is such that

$$\overrightarrow{OM} = \overrightarrow{OO'} + \mathbf{a},$$

which is to say

$$\overrightarrow{O'M} = \mathbf{a}.$$

It is clear from this definition that if the vector sub-space is r-dimensional then the corresponding affine sub-space is an r-dimensional affine space. We denote this by $\mathscr{V}_r$.

As an example, consider the three-dimensional affine point space $\mathscr{E}_3$ of elementary geometry. The one- and two-dimensional affine sub-spaces are then simply the lines and planes of $\mathscr{E}_3$.

29. Euclidean point spaces. DEFINITION: *An affine point space which is associated with a Euclidean vector space is called a Euclidean point space.*

If the associated vector space E_n is properly Euclidean the space $\mathscr{E}_n$ is also said to be properly Euclidean. Furthermore, the reference frame $(O, \mathbf{e}_i)$ is orthonormal if the basis $(\mathbf{e}_i)$ of E_n is orthonormal.

The point space $\mathscr{E}_3$ of elementary geometry is an obvious example of a three-dimensional proper Euclidean point space.

The concept of distance may be introduced through the following definition: The square of the distance between two points A and B in $\mathscr{E}_n$ is the norm of the vector $\overrightarrow{AB}$ (or of $\overrightarrow{BA}$) so that

$$N(\overrightarrow{AB}) = (\overrightarrow{AB})^2 = (AB)^2.$$

If the space is properly Euclidean, $(AB)^2$ is positive definite for two distinct points A, B so that the square root is always real and defines the distance AB between the points A and B. If A, B, C denote any three points in a proper Euclidean space we have

$$\overrightarrow{AC} = \overrightarrow{AB} + \overrightarrow{BC},$$

and applying the inequality (17.3) it follows that

$$AC \leqslant AB + BC. \tag{29.1}$$

This is known as the *triangular inequality*.

Let us return to the general case of a Euclidean point space which we suppose to be referred to an arbitrary frame $(O, \mathbf{e}_i)$. We wish to express analytically the square of the distance between two points. If M has the coordinates (x^i) and N the coordinates (y^i) the vector $\overrightarrow{MN}$ has the components $(y^i - x^i)$. Accordingly, using the notation of Chapter I

$$(MN)^2 = (\overrightarrow{MN})^2 = g_{ij}(y^i - x^i)(y^j - x^j). \tag{29.2}$$

Suppose the point N to be in the immediate neighbourhood of M and let its coordinates be $(x^i + dx^i)$. The square $(ds)^2$ of the distance between M and N is therefore given by the quadratic differential form

$$(ds)^2 = g_{ij} dx^i dx^j, \tag{29.3}$$

where the g_{ij} are constants with respect to the (x^i). It is clear, in particular, that if $\mathscr{E}_n$ is properly Euclidean and is referred to an orthonormal reference system, then

$$(ds)^2 = (dx^1)^2 + (dx^2)^2 + \ldots + (dx^n)^2. \tag{29.4}$$

The reader will recognize this as the generalization to n dimensions of the expression for $(ds)^2$ in terms of orthogonal coordinates in elementary geometry.

CHAPTER III

Tensor Algebra

I. CONCEPT OF A TENSOR PRODUCT

30. Tensor product of two spaces. Let us consider two vector spaces E_n and F_p of n- and p-dimensions respectively and associate with them a vector space of dimensionality np denoted by $E_n \otimes F_p$. If $\mathbf{x}$ and $\mathbf{y}$ belong to E_n and F_p respectively we can set up a correspondence between the pair $(\mathbf{x},\mathbf{y})$ and an element of the vector space $E_n \otimes F_p$ denoted by $\mathbf{x} \otimes \mathbf{y}$. This correspondence has the following properties:

(*a*) If $\mathbf{x}, \mathbf{x}_1, \mathbf{x}_2$ belong to E_n, and $\mathbf{y}, \mathbf{y}_1, \mathbf{y}_2$ to F_p then the distributive law holds with respect to vector addition:

$$\mathbf{x} \otimes (\mathbf{y}_1 + \mathbf{y}_2) = \mathbf{x} \otimes \mathbf{y}_1 + \mathbf{x} \otimes \mathbf{y}_2,$$
$$(\mathbf{x}_1 + \mathbf{x}_2) \otimes \mathbf{y} = \mathbf{x}_1 \otimes \mathbf{y} + \mathbf{x}_2 \otimes \mathbf{y}.$$

(*b*) If α is an arbitrary scalar, the associative law holds:

$$\alpha \mathbf{x} \otimes \mathbf{y} = \mathbf{x} \otimes \alpha \mathbf{y} = \alpha(\mathbf{x} \otimes \mathbf{y}).$$

(*c*) If $(\mathbf{x}_1, \mathbf{x}_2, \ldots, \mathbf{x}_n)$ and $(\mathbf{y}_1, \mathbf{y}_2, \ldots, \mathbf{y}_p)$ are any two bases of E_n and F_p respectively the np elements

$$\mathbf{x}_i \otimes \mathbf{y}_\alpha \quad (i = 1, 2, \ldots, n; \alpha = 1, 2, \ldots, p)$$

of $E_n \otimes F_p$ form a basis in that space.

When these conditions hold we say that the vector space $E_n \otimes F_p$ is the *tensor product* of the vector spaces E_n and F_p and that the element $\mathbf{x} \otimes \mathbf{y}$ is the tensor product of the two vectors $\mathbf{x}$ and $\mathbf{y}$.

31. Analytical expression for the tensor product of two vectors. Let us see how the three properties given above allow the formulation of a rule of composition for the two vectors $\mathbf{x}$ and $\mathbf{y}$. Choose any three bases $(\mathbf{e}_i)$, $(\mathbf{f}_\alpha)$, $(\boldsymbol{\epsilon}_{i\alpha})$ in the vector spaces E_n, F_p and $E_n \otimes F_p$ $(i = 1, 2, \ldots, n; \alpha = 1, 2, \ldots, p)$. According to (*c*) it is possible to write

$$\mathbf{e}_i \otimes \mathbf{f}_\alpha = \boldsymbol{\epsilon}_{i\alpha}. \tag{31.1}$$

Suppose that

$$\left.\begin{array}{l} \mathbf{x} = x^i \mathbf{e}_i, \\ \mathbf{y} = y^\alpha \mathbf{f}_\alpha, \end{array}\right\} \tag{31.2}$$

are two arbitrary vectors belonging respectively to E_n and F_p. Taking the tensor product of the right-hand sides of the equations (31.2) and using properties (a) and (b) we get

$$\mathbf{x} \otimes \mathbf{y} = x^i y^\alpha \mathbf{e}_i \otimes \mathbf{f}_\alpha = x^i y^\alpha \boldsymbol{\epsilon}_{i\alpha}. \tag{31.3}$$

It follows that the quantities $x^i y^\alpha$ are the components of the tensor product $\mathbf{x} \otimes \mathbf{y}$ with respect to the basis $\boldsymbol{\epsilon}_{i\alpha}$.

On the other hand, does the composition rule defined by (31.3) satisfy property (c)? Let $(\mathbf{x}_i)$ and $(\mathbf{y}_\alpha)$ be two arbitrary bases of E_n and F_p; referring the vectors $\mathbf{e}_k$ to the basis $(\mathbf{x}_i)$ we have

$$\mathbf{e}_k = a_k^i \mathbf{x}_i.$$

Similarly

$$\mathbf{f}_\beta = b_\beta^\alpha \mathbf{y}_\alpha.$$

Each element T of $E_n \otimes F_p$ can be written

$$\mathbf{T} = t^{i\alpha} \boldsymbol{\epsilon}_{i\alpha} = t^{i\alpha} \mathbf{e}_i \otimes \mathbf{f}_\alpha \tag{31.4}$$

so that

$$\mathbf{T} = t^{k\beta} a_k^i \mathbf{x}_i \otimes b_\beta^\alpha \mathbf{y}_\alpha = t^{k\beta} a_k^i b_\beta^\alpha \mathbf{x}_i \otimes \mathbf{y}_\alpha.$$

Therefore $\mathbf{T}$ can be expressed as a linear combination of the elements $\mathbf{x}_i \otimes \mathbf{y}_\alpha$. If $\mathbf{T} = 0$, equation (31.4) shows that $t^{i\alpha} = 0$, and the system of np elements $\mathbf{x}_i \otimes \mathbf{y}_\alpha$ is linearly independent, which shows that (c) is satisfied. This can be expressed formally as follows:

THEOREM: *If the spaces E_n, F_p, $E_n \otimes F_p$ are referred to certain bases related according to (31.1) then the only composition rule satisfying the properties of §30 is that which, for a vector $\mathbf{x}$ with components x^i and a vector $\mathbf{y}$ with components y^α, gives a correspondence with the element of $E_n \otimes F_p$ having components $x^i y^\alpha$.*

32. Tensor products of several spaces. Tensors. Consider three vector spaces E_n, F_p, G_q with n, p, q dimensions respectively. If $\mathbf{x}$ belongs to E_n, $\mathbf{y}$ to F_p, $\mathbf{z}$ to G_q then the element $\mathbf{x} \otimes \mathbf{y}$ of $E_n \otimes F_p$ can be multiplied tensorially with the element $\mathbf{z}$ of G_q. The element $[\mathbf{x} \otimes \mathbf{y}] \otimes \mathbf{z}$ of a vector space H is thus obtained. We assume that the same element of H is obtained by taking the tensor product of $\mathbf{x}$ with $\mathbf{y} \otimes \mathbf{z}$, i.e.

$$[\mathbf{x} \otimes \mathbf{y}] \otimes \mathbf{z} = \mathbf{x} \otimes [\mathbf{y} \otimes \mathbf{z}] \qquad (32.1)$$

which is the associative property of tensor products. It is only necessary to assume this property for the basis vectors, then (32.1) follows because of the distributive properties (*a*) and (*b*) above. We denote the common value of the two sides of (32.1) by $\mathbf{x} \otimes \mathbf{y} \otimes \mathbf{z}$ and the vector space H will be represented by $E_n \otimes F_p \otimes G_q$.

Given a finite number r of vector spaces $E_n, F_p, G_q, \ldots$, the definition of their tensor product follows immediately. As every element of $E_n \otimes F_p \otimes G_q \otimes \ldots$ is not necessarily the tensor product of r vectors belonging respectively to $E_n, F_p, G_q, \ldots$† we formulate the following general definition:

DEFINITION: *Each element of the vector space*

$$E_n \otimes F_p \otimes G_q \otimes \ldots$$

constructed from the spaces E_n, F_p, G_q is called a tensor.

II. AFFINE TENSORS

33. Affine tensors attached to a vector space. Given an n-dimensional space E_n it is possible to construct the tensor product of q spaces identical with E_n. The vector space of n^q dimensions thus obtained will be denoted by $E_n^{(q)}$ and called the qth tensorial power of E_n. As the vector spaces E_n and $E_n^{(q)}$ are referred respectively to the basis vectors $(\mathbf{e}_i)$ and $(\mathbf{e}_{i_1 i_2 \cdots i_q})$ we adopt the following convention which generalizes (31.1)

$$\mathbf{e}_{i_1} \otimes \mathbf{e}_{i_2} \otimes \mathbf{e}_{i_3} \ldots \otimes \mathbf{e}_{i_q} = \mathbf{e}_{i_1 i_2 \ldots i_q}.$$

If $\mathbf{x}_{(1)}, \mathbf{x}_{(2)}, \ldots, \mathbf{x}_{(q)}$ denote any q vectors of E_n with components $x^{i_1}_{(1)}, x^{i_2}_{(2)}, \ldots, x^{i_q}_{(q)}$ respectively, then

$$\mathbf{x}_{(1)} \otimes \mathbf{x}_{(2)} \otimes \ldots \otimes \mathbf{x}_{(q)} = x^{i_1}_{(1)} x^{i_2}_{(2)} \ldots x^{i_q}_{(q)} \mathbf{e}_{i_1 i_2 \ldots i_q}. \qquad (33.1)$$

† It may be the sum of several such expressions. (T.)

Consider the vector space E_n^* dual to E_n referred to the basis (x^i) dual to (e_i). Starting from these two spaces we can carry out the following operation: take arbitrary tensor powers of E_n or E_n^* and multiply them tensorially amongst themselves. We then obtain tensor products of the type

$$E_n^{(r_1)} \otimes E_n^{*(s_1)} \otimes E_n^{(r_2)} \otimes \ldots \otimes E_n^{*(s_p)} \otimes E_n^{(r_m)}. \qquad (33.2)$$

DEFINITION: *A tensor is said to be an affine tensor attached to the vector space E_n if it is an element of a vector space obtained by forming a vector product of spaces identical with E_n or its dual.*

If $(r_1 + r_2 + \ldots + r_m) + (s_1 + s_2 + \ldots + s_p) = q$, the vector space (33.2) is q dimensional. Each element of (33.2) is thus called an affine tensor of *order*† q; it is $(r_1 + r_2 + \ldots + r_m)$-fold contravariant and $(s_1 + s_2 + \ldots + s_p)$-fold covariant. Each element of $E_n^{(q)}$ is a contravariant affine tensor of order q and each element of $E_n^{*(q)}$ is a covariant affine tensor of order q. In particular it is often said that the elements of E_n are contravariant vectors whilst those of E_n^* are covariant vectors. Naturally this terminology varies according to which of the spaces E_n and E_n^* is considered to be given first.

34. Components of an affine tensor. In order to represent the components of different affine tensors it is convenient to introduce a special notation which will now be explained. We consider, for example, a tensor of order q which is an element of the space $E_n^{(q-2)} \otimes E_n^{*(2)}$. A basis of this vector space is formed by the set of elements

$$\mathbf{e}_{i_1} \otimes \mathbf{e}_{i_2} \otimes \ldots \otimes \mathbf{e}_{i_{(q-2)}} \otimes x^{i_{(q-1)}} \otimes x^{i_q} = \mathbf{\epsilon}_{i_1 i_2 \ldots i_{q-2}}{}^{i_{q-1} i_q}, \qquad (34.1)$$

where both sides have the same suffices and superfixes. Let $\mathbf{T}$ be a tensor element of $E_n^{(q-2)} \otimes E_n^{*(2)}$ and refer it to the basis (34.1). In order to preserve the summation convention we represent the components of $\mathbf{T}$ by $t^{i_1 i_2 \ldots i_{q-2}}{}_{i_{q-1} i_q}$ so that

$$\mathbf{T} = t^{i_1 i_2 \ldots i_{q-2}}{}_{i_{q-1} i_q} \mathbf{\epsilon}_{i_1 i_2 \ldots i_{q-2}}{}^{i_{q-1} i_q}. \qquad (34.2)$$

The indices $i_1, i_2, \ldots i_{q-2}$ are said to be contravariant and the i_{q-1}, i_q are said to be covariant. Mere inspection of the components of a

† The term *rank* is often used in this context. (T.)

tensor is therefore sufficient to determine how many times it is contra-variant and covariant. Using (34.1) and (34.2) we deduce

$$\mathbf{T} = \sum_{i_2\ldots i_q} (\sum_{i_1} t^{i_1 i_2\ldots i_{q-2}}{}_{i_{q-1} i_q} \mathbf{e}_{i_1}) \otimes \mathbf{e}_{i_2} \otimes \ldots \otimes \mathbf{e}_{i_{q-2}} \otimes x^{i_{q-1}} \otimes x^{i_q}$$

where, for clarity, the summation signs have been inserted.

Each term of the summation on the right-hand side is the tensor product of q contravariant or covariant vectors and this sum has n^{q-1} terms.

Thus every affine tensor of order q is the sum of at most n^{q-1} tensor products of q contravariant or covariant vectors.

For example, if the tensor $\mathbf{T}$, an element of $E_n^{(q)}$, is the sum of $p < n^{q-1}$ tensor products of q vectors it can be expressed as the sum of p terms in the form of the right-hand side of (33.1) and its components will be the sum of p terms in the form

$$x^{i_1}_{(1)} x^{i_2}_{(2)} \ldots x^{i_q}_{(q)}.$$

It is clear that, whatever the type of tensor, all its components have the same form.

35. Change of basis for the components of an affine tensor. Let us again consider a tensor $\mathbf{T}$ which is an element of the space $E_n^{(q-2)} \otimes E_n^{*(2)}$. If the space E_n is referred to the basis $(\mathbf{e}_i)$, E_n^* will be referred to the dual basis (x^i) and $E_n^{(q-2)} \otimes E_n^{*(2)}$ to the basis

$$\mathbf{e}_{i_1} \otimes \ldots \otimes \mathbf{e}_{i_{(q-2)}} \otimes x^{i_{q-1}} \otimes x^{i_q}. \tag{35.1}$$

Now refer E_n to another basis $(\mathbf{e}_{j'})$ defined by

$$\mathbf{e}_i = A_i^{j'} \mathbf{e}_{j'}, \qquad \mathbf{e}_{j'} = A_{j'}^{i} \mathbf{e}_i. \tag{35.2}$$

The space $E_n^{(q-2)} \otimes E_n^{*(2)}$ will then be referred to the associated basis

$$\mathbf{e}_{j'_1} \otimes \mathbf{e}_{j'_2} \otimes \ldots \otimes \mathbf{e}_{j'_{q-2}} \otimes x^{j'_{q-1}} \otimes x^{j'_q}. \tag{35.3}$$

We wish to determine the components $t^{i_1 i_2 \ldots i_{q-2}}{}_{i_{q-1} i_q}$ of $\mathbf{T}$ with respect to the first basis in terms of the components $t^{j'_1 j'_2 \ldots j'_{q-2}}{}_{j'_{q-1} j'_q}$ referred to the second.

To this end suppose that **T** is the tensor product of q contravariant or covariant vectors. Then

$$t^{i_1 i_2 \ldots i_{q-2}}{}_{i_{q-1} i_q} = x^{i_1}_{(1)} x^{i_2}_{(2)} \ldots x^{i_{q-2}}_{(q-2)} x_{(q-1) i_{q-1}} x_{(q) i_q}; \quad (35.4)$$

$$t^{j'_1 j'_2 \ldots j'_{q-2}}{}_{j'_{q-1} j'_q} = x^{j'_1}_{(1)} x^{j'_2}_{(2)} \ldots x^{j'_{q-2}}_{(q-2)} x_{(q-1) j'_{q-1}} x_{(q) j'_q}. \quad (35.5)$$

However, using the transformation formulae for the components of a contravariant or covariant vector, we have

$$x^{i_1}_{(1)} = A^{i_1}_{j'_1} x^{j'_1}_{(1)} \text{etc.}; \qquad x_{(q-1) i_{q-1}} = A^{j'_{q-1}}_{i_{q-1}} x_{(q-1) j'_{q-1}} \text{etc.}$$

Hence

$$t^{i_1 i_2 \ldots i_{q-2}}{}_{i_{q-1} i_q} = A^{i_1}_{j'_1} \ldots A^{j'_q}_{i_q} t^{j'_1 j'_2 \ldots j'_{q-2}}{}_{j'_{q-1} j'_q} \quad (35.6)$$

or, interchanging the roles of the two bases we obtain

$$t^{j'_1 j'_2 \ldots j'_{q-2}}{}_{j'_{q-1} j'_q} = A^{j'_1}_{i_1} \ldots A^{i_q}_{j'_q} t^{i_1 i_2 \ldots i_{q-2}}{}_{i_{q-1} i_q}. \quad (35.7)$$

As each tensor is to be regarded as the sum of p tensor products and the relations (35.6) and (35.7) hold for each component of **T**, these relations also hold for the components of an arbitrary affine tensor which is $(q-2)$-fold contravariant and two-fold covariant.

Using (35.6) and (35.7) the general transformation rule for the components of any tensor may easily be derived. It is clear that a repetition of the above reasoning will, in particular, give the following expressions for the components of a purely contravariant tensor:

$$t^{i_1 i_2 \ldots i_q} = A^{i_1}_{j'_1} \ldots A^{i_q}_{j'_q} t^{j'_1 j'_2 \ldots j'_q}; \quad (35.8)$$

$$t^{j'_1 j'_2 \ldots j'_q} = A^{j'_1}_{i_1} \ldots A^{j'_q}_{i_q} t^{i_1 i_2 \ldots i_q}. \quad (35.9)$$

The components of a covariant tensor transform according to

$$t_{i_1 \ldots i_q} = A^{j'_1}_{i_1} \ldots A^{j'_q}_{i_q} t_{j'_1 \ldots j'_q}; \quad (35.10)$$

$$t_{j'_1 \ldots j'_q} = A^{i_1}_{j'_1} \ldots A^{i_q}_{j'_q} t_{i_1 \ldots i_q}. \quad (35.11)$$

These transformation formulae can be interpreted as follows. Let E_n and $E_n^{(q)}$ be referred to associated bases $(\mathbf{e}_i)$ and $(\mathbf{e}_{i_1} \otimes \ldots \otimes \mathbf{e}_{i_q})$ and consider a system of n^q quantities $t^{i_1 \cdots i_q}$ which transform according to (35.8) and (35.9) on passing from the bases $(\mathbf{e}_i)$, $(\mathbf{e}_{i_1} \otimes \mathbf{e}_{i_2} \otimes \ldots \otimes \mathbf{e}_{i_q})$ to $(\mathbf{e}_{j'})$, $(\mathbf{e}_{j'_1} \otimes \ldots \otimes \mathbf{e}_{j'_q})$. A correspondence can be set up between these quantities and a tensor **T** which has them as components with respect to the basis $(\mathbf{e}_{i_1} \otimes \ldots \otimes \mathbf{e}_{i_q})$. The transformation rule then defines the same tensor **T** with respect to an arbitrary basis

$$(\mathbf{e}_{i'_1} \otimes \ldots \otimes \mathbf{e}_{i'_q}).$$

THEOREM: *In order that the n^q quantities $t^{i_1 i_2 \ldots i_q}$ associated with a basis $(\mathbf{e}_{i_1} \otimes \ldots \otimes \mathbf{e}_{i_q})$ of the space $E_n^{(q)}$ can be considered as the components of some contravariant tensor it is necessary and sufficient that the system transforms according to (35.8) and (35.9) under a change of basis.*

Similar statements hold for the components of a mixed tensor of any kind.

36. A criterion of tensor character. It is easy to deduce from the preceding results a criterion of tensor character which is of great practical use. By way of a change we shall consider the case of a purely covariant tensor.

THEOREM: *A necessary and sufficient condition for a system of n^q quantities $t_{i_1 \ldots i_q}$ referred to a basis $(x^{i_1} \otimes \ldots \otimes x^{i_q})$ of the space $E_n^{*(q)}$ to be the components of a covariant tensor is that for any contravariant vectors $[\mathbf{x}_{(1)}, \mathbf{x}_{(2)} \ldots \mathbf{x}_{(q)}]$ with components $x^i_{(j)}$ the quantity*

$$t_{i_1 i_2 \ldots i_q} x^{i_1}_{(1)} x^{i_2}_{(2)} \ldots x^{i_q}_{(q)}$$

shall remain invariant with respect to all changes of basis.

We establish first that the condition is necessary. If the $t_{i_1 i_2 \ldots i_q}$ are the components of a covariant tensor they transform according to

$$t_{i_1 i_2 \ldots i_q} = A^{j'_1}_{i_1} \ldots A^{j'_q}_{i_q} t_{j'_1 j'_2 \ldots j'_q}$$

and the components of the q given contravariant vectors transform according to

$$x^{i_1}_{(1)} = A^{i_1}_{k'_1} x^{k'_1}_{(1)}; \quad \ldots; \quad x^{i_q}_{(q)} = A^{i_q}_{k'_q} x^{k'_q}_{(q)}.$$

Hence we deduce

$$t_{i_1 i_2 \ldots i_q} x^{i_1}_{(1)} x^{i_2}_{(2)} \ldots x^{i_q}_{(q)} = A^{j'_1}_{i_1} A^{i_1}_{k'_1} \ldots t_{j'_1 j'_2 \ldots j'_q} x^{k'_1}_{(1)} \ldots x^{k'_q}_{(q)}.$$

Since $A_i^{j'} A_{k'}^i$ represents the components of the $e_{k'}$ referred to the $e_{j'}$ we have

$$A_i^{j'} A_{k'}^i = \delta_{k'}^{j'} = \begin{cases} 0 & \text{if } j' \neq k' \\ 1 & \text{if } j' = k'. \end{cases} \tag{36.1}$$

It follows that

$$t_{i_1 \ldots i_q} x_{(1)}^{i_1} \ldots x_{(q)}^{i_q} = t_{j'_1 \ldots j'_q} x_{(1)}^{j'_1} \ldots x_{(q)}^{j'_q}. \tag{36.2}$$

Conversely, suppose the equality (36.2) to be satisfied for the arbitrary set of contravariant vectors $[\mathbf{x}_{(1)} \ldots \mathbf{x}_{(q)}]$. Then the components $x_{(k)}^{j'}$ of these vectors transform according to the formulae

$$x_{(1)}^{j'_1} = A_{i_1}^{j'_1} x_{(1)}^{i_1}; \quad \ldots; \quad x_{(q)}^{j'_q} = A_{i_q}^{j'_q} x_{(q)}^{i_q}.$$

We therefore have the relation

$$t_{i_1 \ldots i_q} x_{(1)}^{i_1} \ldots x_{(q)}^{i_q} = A_{i_1}^{j'_1} \ldots A_{i_q}^{j'_q} t_{j'_1 \ldots j'_q} x_{(1)}^{i_1} \ldots x_{(q)}^{i_q}.$$

It follows that the $t_{i_1 \ldots i_q}$ transform according to (35.10). This completes the required proof.

The above theorem can immediately be generalized as follows:

THEOREM: *In order that a system of n^{p+q} quantities $t_{i_1 \ldots i_{p+q}}$ referred to the basis $(x^{i_1} \otimes \ldots \otimes x^{i_p} \otimes \ldots \otimes x^{i_{p+q}})$ of the space $E_n^{*(p+q)}$ can be identified with the components of a covariant tensor, it is necessary and sufficient that, for p arbitrary contravariant vectors $[\mathbf{x}_{(1)}, \mathbf{x}_{(2)}, \ldots, \mathbf{x}_{(p)}]$ with components $x_{(J)}^i$, the quantities*

$$t_{i_1 i_2 \ldots i_p i_{p+1} \ldots i_{p+q}} x_{(1)}^{i_1} \ldots x_{(p)}^{i_p}$$

shall be the components of a covariant tensor of order q.

Let $(\mathbf{y}_{(1)}, \mathbf{y}_{(2)}, \ldots, \mathbf{y}_{(q)})$ be q arbitrary contravariant vectors with components $y_{(J)}^i$. In order that the quantities $t_{i_1 \ldots i_p i_{p+1} \ldots i_{p+q}}$ shall be the components of a covariant tensor of order $p+q$ it is necessary and sufficient that the quantity

$$t_{i_1 \ldots i_p i_{p+1} \ldots i_{p+q}} x_{(1)}^{i_1} \ldots x_{(p)}^{i_p} y_{(1)}^{i_{p+1}} \ldots y_{(q)}^{i_{p+q}}$$

shall remain invariant with respect to changes of basis. But this condition is also the necessary and sufficient condition that the quantities

$$t_{i_1 \ldots i_p i_{p+1} \ldots i_{p+q}} x_{(1)}^{i_1} \ldots x_{(p)}^{i_p}$$

shall be the components of a covariant tensor of order q.

Analogous statements can be made for the components of an affine tensor of any kind.

37. Affine tensor algebra. We have already met, incidentally, some algebraic operations which permit the formation of new tensors from known ones. Let us revue these briefly:

(*a*) *Tensor addition.* Given two tensors of the same order and of the same kind which are both elements, for example, of $E_n^{(q-2)} \otimes E_n^{*(2)}$, the procedure of vector addition generates a third tensor of order q and of the same kind which is called their sum. If the two tensors under consideration have the components $t^{i_1 \ldots i_{q-2}}{}_{i_{q-1} i_q}$ and $u^{i_1 \ldots i_{q-2}}{}_{i_{q-1} i_q}$ respectively their sum obviously has the components

$$s^{i_1 \ldots i_{q-2}}{}_{i_{q-1} i_q} = t^{i_1 \ldots i_{q-2}}{}_{i_{q-1} i_q} + u^{i_1 \ldots i_{q-2}}{}_{i_{q-1} i_q}.$$

(*b*) *Tensor multiplication.* Given two tensors of order q and q' which are of any kind, their tensor product generates a tensor of order $q+q'$. If, for example, two tensors have the components $t^{i_1 \ldots i_{q-1}}{}_{i_q}$ and $u_{i_{q+1}}{}^{i_{q+2} \ldots i_{q+q'}}$, then their tensor product has components

$$p^{i_1 \ldots i_{q-1}}{}_{i_q i_{q+1}}{}^{i_{q+2} \ldots i_{q+q'}} = t^{i_1 \ldots i_{q-1}}{}_{i_q} \cdot u_{i_{q+1}}{}^{i_{q+2} \ldots i_{q+q'}}.$$

Consider any tensor which is an element, for example, of $E_n^{(q)}$. The multiplication in $E_n^{(q)}$ of this tensor by a scalar is a particular case of tensor multiplication if the scalar (or invariant) is regarded as a tensor of zero order. We shall adopt this point of view in the future.

38. Contraction of indices. Besides the two fundamental operations given above there exists a third, the contraction of indices, which permits the derivation of new tensors of order $(q-2)$ from a given mixed tensor of order q.

Consider first a mixed tensor of order 2 having the components $t_i{}^j$. We now show that the quantity $t_i{}^i$, obtained by summing the components with identical contravariant and covariant indices, is invariant with respect to a change in basis.

In fact, on changing the basis we get

$$t_i{}^i = A_i^{h'} A_{k'}^i t_{h'}{}^{k'}$$

But, from (35.1)

$$A_i^{h'} A_{k'}^i = \delta_{k'}^{h'} = \begin{cases} 0 & \text{if } k' \neq h' \\ 1 & \text{if } k' = h', \end{cases}$$

which gives

$$t_i^{\ i} = t_{k'}^{\ k'}.$$

Consider next a mixed tensor of order q and choose a pair of indices, one covariant and one contravariant. In order to simplify the notation let these be the first two indices of the tensor

$$t_{i_1}^{\ i_2}{}_{i_3 \ldots i_q}.$$

Let i_1 and i_2 be put equal to i and sum over this repeated index. We now show that the quantities

$$t_i^{\ i}{}_{i_3 \ldots i_q}$$

thus obtained are the components of a tensor of order $(q-2)$. Let $[\mathbf{x}_{(3)}, \ldots, \mathbf{x}_{(q)}]$ be $(q-2)$ arbitrary contravariant vectors. In virtue of the results of §36 the quantities

$$t_{i_1}^{\ i_2}{}_{i_3 \ldots i_q} x_{(3)}^{i_3} \ldots x_{(q)}^{i_q}$$

are the components of a mixed second-order tensor. It follows that for any vectors $[\mathbf{x}_{(3)} \ldots \mathbf{x}_{(q)}]$ the quantity

$$t_i^{\ i}{}_{i_3 \ldots i_q} x_{(3)}^{i_3} \ldots x_{(q)}^{i_q}$$

is invariant with respect to changes of basis which demonstrates the tensor character of $t_i^{\ i}{}_{i_3 \ldots i_q}$.

The operation which consists in equating two indices, one contravariant and the other covariant, and summing over their common values is called contraction. The contraction of two indices in a tensor of order q generates a tensor of order $(q-2)$.

It is clear that if the tensor under consideration has more than one pair of indices, one covariant and the other contravariant, the operation of contraction can be repeated.

39. Contracted multiplication. A general criterion of tensor character.
One frequently encounters tensor products where the contracted indices belong to different factors of the product. We shall call this contracted multiplication. The operation of contraction may, moreover, be repeated many times in such circumstances.

If t_{ijkl} and u^{mnr} are the components of two tensors their tensor product has the components

$$p_{ijkl}{}^{mnr} = t_{ijkl}u^{mnr},$$

and the tensor

$$p_{ijkl}{}^{klr} = t_{ijkl}u^{klr}$$

is one of their contracted products, obtained by contracting the indices m, k and n, l.

The concept of contracted multiplication can be used to specify a criterion of tensor character which generalizes that of §36. This can best be expressed in terms of a particular example:

In order that the quantities t^{ijkl} referred to a basis $(\mathbf{e}_i \otimes \mathbf{e}_j \otimes \mathbf{e}_k \otimes \mathbf{e}_l)$ shall be the components of a contravariant tensor it is necessary and sufficient that, for any covariant tensor s_{kl}, the quantities $t^{ijkl}s_{kl}$ shall be the components of a contravariant tensor.

The condition is evidently necessary from the study of contracted multiplication which we have just carried out. To show that it is sufficient we only have to observe that it is possible for s_{kl} to be the tensor product of two covariant vectors with components x_k and y_l, and to use the second theorem of §36.

40. Symmetric and antisymmetric tensors. A contravariant second-order tensor, whose components referred to a given basis are t^{ij}, is said to be *symmetric* in its two indices if

$$t^{ij} = t^{ji}$$

and *antisymmetric* if $\qquad t^{ij} = -t^{ji}.$

Suppose, for example, that it is symmetric and change the basis:

$$t^{k'l'} = A_i^{k'} A_j^{l'} t^{ij} = A_i^{k'} A_j^{l'} t^{ji} = t^{l'k'}.$$

Hence the fact that a tensor is symmetric (or antisymmetric) is a property of the tensor itself and not merely of its components with respect to a particular basis.

The same considerations hold for a covariant tensor of second order. They extend immediately to pairs of indices, both contravariant or both covariant, of tensors of order $q > 2$.

III. EUCLIDEAN TENSORS

41. Euclidean tensors and their components. Let us now take E_n to be a *Euclidean* vector space. In this case we shall show that every tensor of order q can be identified with a contravariant tensor of order q, so there is no point in considering contravariant, covariant or mixed tensors of the same order to be distinct entities.

Suppose E_n to be referred to an arbitrary basis and E_n^* to the corresponding dual basis and designate the coefficients of the fundamental quadratic form by g_{ij}. We have seen in §24 that we may identify the element in E_n which has the components x^i with the element in E_n^* which has the components x_i since

$$x_i = g_{ij}x^j, \qquad x^i = g^{ij}x_j.$$

Having established this, let us consider a contravariant tensor **T** of order q which is the *tensor product* of q vectors $\mathbf{x}_{(1)}, \mathbf{x}_{(2)}, \ldots, \mathbf{x}_{(q)}$:

$$\mathbf{T} = \mathbf{x}_{(1)} \otimes \mathbf{x}_{(2)} \otimes \ldots \otimes \mathbf{x}_{(q)}.$$

To each of these vectors there corresponds a particular element in E_n^*. Consider the various affine tensors obtained by replacing one or more of the vectors $\mathbf{x}_{(1)}, \mathbf{x}_{(2)}, \ldots, \mathbf{x}_{(q)}$ by their corresponding elements in E_n^*. These affine tensors are said to define the *same Euclidean tensor* whose contravariant, covariant or mixed components are the components of the corresponding affine contravariant, covariant or mixed tensors.

The relations between the various types of component of the Euclidean tensor **T** may readily be determined. Denoting the contravariant components of the vector $\mathbf{x}_{(1)}$ by $x^i_{(1)}$ the contravariant components of **T** are

$$t^{i_1 i_2 \ldots i_q} = x^{i_1}_{(1)} x^{i_2}_{(2)} \ldots x^{i_q}_{(q)}.$$

If, for example, we replace the vector $\mathbf{x}_{(2)}$ by the corresponding element of E_n^* we obtain the mixed components, one-fold covariant:

$$t^{i_1}{}_{i_2}{}^{i_3 \ldots i_q} = x^{i_1}_{(1)} x_{(2)i_2} x^{i_3}_{(3)} \ldots x^{i_q}_{(q)},$$

where $$x_{(2)i_2} = g_{i_2 j_2} x^{j_2}_{(2)}.$$

From this we deduce that

$$t^{i_1}{}_{i_2}{}^{i_3\ldots i_q} = g_{i_2 j_2} t^{i_1 j_2 i_3 \ldots i_q}. \tag{41.1}$$

Conversely, we have

$$t^{i_1 \ldots i_q} = g^{i_2 j_2} t^{i_1}{}_{j_2}{}^{i_3 \ldots i_q}. \tag{41.2}$$

If this operation is repeated using the index i_3 we obtain the two-fold covariant components

$$t^{i_1}{}_{i_2 i_3}{}^{i_4 \ldots i_q} = g_{i_3 j_3} t^{i_1}{}_{i_2}{}^{j_3 i_4 \ldots i_q} = g_{i_2 j_2} g_{i_3 j_3} t^{i_1 j_2 j_3 i_4 \ldots i_q}. \tag{41.3}$$

Operating on all the indices we finally obtain for the covariant components

$$t_{i_1 i_2 \ldots i_q} = g_{i_1 j_1} \cdots g_{i_q j_q} t^{j_1 \ldots j_q}. \tag{41.4}$$

Conversely, the contravariant components may be expressed in terms of the covariant components as

$$t^{i_1 \ldots i_q} = g^{i_1 j_1} \cdots g^{i_q j_q} t_{j_1 \ldots j_q}. \tag{41.5}$$

It is now seen that, on multiplying by g_{ij} or g^{ij} and summing, each of the q indices of the tensor **T** may be placed either in the contravariant or in the covariant position.

Let **T** now be any contravariant tensor of order q; such a tensor can be expressed as a sum of p tensor products of q vectors. Each of these tensor products defines a Euclidean tensor. Consider the various affine tensors of the same type associated with these Euclidean tensors. In summing them one sets up a correspondence between the tensor **T** and the affine tensors whose components are given by formulae such as (41.1), (41.3), (41.4) by virtue of the linear character of these formulae in the components $t^{i_1 i_2 \ldots i_q}$ of **T**. It follows that these affine tensors do not depend on the way in which **T** has been decomposed into a sum of tensor products and they are consequently uniquely defined by **T**. We expect these various affine tensors to define a unique Euclidean tensor whose various components are the components of identical affine tensors. This can be expressed formally as follows:

THEOREM: *The various contravariant, covariant or mixed components of a Euclidean tensor can be derived from each other by multiplying by g_{ij} or g^{ij} and summing, this operation being performed one or more times.*

The criteria of tensor character which have already been given for affine tensors obviously apply without modification to Euclidean tensors; the covariant components (for example) of a Euclidean tensor being the same as the components of the corresponding covariant affine tensor.

42. Symmetric and antisymmetric Euclidean tensors. A second-order Euclidean tensor is said to be *symmetric* if the associated contravariant affine tensor is symmetric. We then have

$$t^{ij} = t^{ji}.$$

Therefore

$$t_{kl} = g_{ki}g_{lj}t^{ij} = g_{ki}g_{lj}t^{ji} = t_{lk},$$

and the covariant components are also symmetric in their two indices. Conversely, if the covariant components are symmetric, it is clear that the contravariant components are also symmetric. Analogous considerations hold for antisymmetric Euclidean tensors. They may also be applied to any two indices of a tensor of order $q > 2$.

43. The fundamental tensor. The scalar product of two vectors **x** and **y** of E_n, which have the contravariant components x^i and y^i respectively is given by

$$\mathbf{x} \cdot \mathbf{y} = g_{ij}x^i y^j.$$

This expression is invariant under changes of basis for any pair of vectors **x** and **y**. It therefore follows that the g_{ij} are the covariant components of a Euclidean tensor. We call this the *fundamental tensor* of the space E_n.

This tensor, being symmetric, has only one set of mixed components

$$g_j{}^i = g_{jk}g^{ki} = g^{ik}g_{kj} = g^i{}_j.$$

Let us evaluate these. From the definition of the g^{jk} (21.5) we have

$$g^j{}_i = \frac{\alpha^{jk}g_{kl}}{g},$$

where α^{jk} denotes the cofactor of g_{jk} in the determinant g. The numerator of the right-hand side is therefore the same as the expansion of g except that the g_{kj} are replaced by the g_{ki}. It follows from the properties of such expansions that

$$g^j{}_i = \delta^j{}_i = \begin{cases} 0 & \text{if } i \neq j, \\ 1 & \text{if } i = j.\dagger \end{cases} \tag{43.1}$$

The contravariant components of the fundamental tensor are therefore given by

$$g^{ik} g^j{}_k = g^{ij}.$$

These components are, of course, identical with the quantities g^{ij} introduced in §21. We can now state:

THEOREM: *The quantities g_{ij} and g^{ij} are respectively the covariant and contravariant components of a certain symmetric tensor which is the fundamental tensor of Euclidean space. The quantities defined by (43.1) are the mixed components of this tensor.*

44. Euclidean tensor algebra. We have seen that the various components of a Euclidean tensor may be expressed as linear forms in the contravariant components (for example) of this tensor. It follows that the algebraic operations given for affine tensors may be extended to Euclidean tensors.

(*a*) *Addition.* Tensor addition (considered in §37) gives a correspondence between two Euclidean tensors of the same order q and a third Euclidean tensor of order q called their sum. If the two tensors have contravariant components $t^{i_1 i_2 \ldots i_q}$ and $u^{i_1 \ldots i_q}$, then the contravariant components of their sum are given by

$$s^{i_1 i_2 \ldots i_q} = t^{i_1 i_2 \ldots i_q} + u^{i_1 i_2 \ldots i_q}.$$

Repeated contracted multiplication by the fundamental tensor gives a similar relation for the covariant components:

$$s_{i_1 i_2 \ldots i_q} = t_{i_1 \ldots i_q} + u_{i_1 \ldots i_q}.$$

An analogous result holds for the various components with mixed indices.

† When $i \neq j$ the expansion is that of a determinant with a repeated column, and therefore vanishes. When $i = j$ the expansion is of g.

(b) *Multiplication.* Tensor multiplication generates a Euclidean tensor of order $q+q'$ from two Euclidean tensors of orders q and q'. If the tensors under consideration have contravariant components $t^{i_1 \ldots i_q}$ and $u^{i_{q+1} \ldots i_{q+q'}}$, the contravariant components of their product are given by

$$p^{i_1 \ldots i_q i_{q+1} \ldots i_{q+q'}} = t^{i_1 \ldots i_q} u^{i_{q+1} \ldots i_{q+q'}}.$$

This equality obviously remains valid if one or more of the similarly positioned indices of the two sides is placed in the covariant position.

(c) *Contraction.* Consider a Euclidean tensor with contravariant components $t^{i_1 i_2 \ldots i_q}$ and choose two arbitrary indices, for example the indices i_1 and i_2. When the first of these is moved to the covariant position the corresponding components are

$$t_{j_1}{}^{i_2 \ldots i_q} = g_{i_1 j_1} t^{i_1 \ldots i_q}.$$

By contracting the indices j_1 and i_2 it follows that

$$t_j{}^{j i_3 \ldots i_q} = g_{ij} t^{iji_3 \ldots i_q} = t^j{}_j{}^{i_3 \ldots i_q}, \tag{44.1}$$

using the symmetry of the g_{ij}. The quantities (44.1) are the contravariant components of a Euclidean tensor of order $(q-2)$:

$$c^{i_3 \ldots i_q} = t_j{}^{j i_3 \ldots i_q}. \tag{44.2}$$

It follows from (44.1) that this tensor does not depend upon which of the indices i_1 and i_2 is lowered. (44.2) remains valid if one or more of the indices $i_3 \ldots i_q$ is placed in the covariant position.

45. The space $E_n^{(q)}$ as a Euclidean space.

A Euclidean space may readily be constructed from $E_n^{(q)}$, which is the qth tensorial power of the Euclidean space E_n.

We refer the space E_n to an arbitrary basis $(\mathbf{e}_i)$ and $E_n^{(q)}$ to the associated basis $(\mathbf{e}_{i_1} \otimes \mathbf{e}_{i_2} \otimes \ldots \otimes \mathbf{e}_{i_q})$. Any two qth order tensors, $\mathbf{T}$ with components $t^{i_1 i_2 \ldots i_q}$ and $\mathbf{U}$ with components $u^{i_1 i_2 \ldots i_q}$, define a scalar

$$\mathbf{T U} = t^{i_1 i_2 \ldots i_q} u_{i_1 i_2 \ldots i_q} \tag{45.1}$$

such that

$$\mathbf{T U} = g_{i_1 j_1} \ldots g_{i_q j_q} t^{i_1 \ldots i_q} u^{j_1 \ldots j_q} = t_{j_1 \ldots j_q} u^{j_1 \ldots j_q}. \tag{45.2}$$

It is clear that the rule of composition thus defined has the properties detailed in §15 which characterize the scalar product of two vectors in a vector space. We say that (45.1) is the scalar product of two tensors **T** and **U**. This scalar product defines a Euclidean space constructed on $E_n^{(q)}$.

In particular the scalar product of two elements $(\mathbf{e}_{i_1} \otimes \mathbf{e}_{i_2} \otimes \ldots \otimes \mathbf{e}_{i_q})$ and $(\mathbf{e}_{j_1} \otimes \mathbf{e}_{j_2} \ldots \otimes \mathbf{e}_{j_q})$ of the base of $E_n^{(q)}$ is according to (45.2), given by

$$(\mathbf{e}_{i_1} \otimes \mathbf{e}_{i_2} \otimes \ldots \otimes \mathbf{e}_{i_q})(\mathbf{e}_{j_1} \otimes \mathbf{e}_{j_2} \otimes \ldots \otimes \mathbf{e}_{j_q}) = g_{i_1 j_1} g_{i_2 j_2} \ldots g_{i_q j_q}, \quad (45.3)$$

so that the fundamental quadratic form of the Euclidean space $E_n^{(q)}$ has the coefficients $g_{i_1 j_1} g_{i_2 j_2} \ldots g_{i_q j_q}$.

If the basis $(\mathbf{e}_i)$ of E_n is orthonormal then

$$(\mathbf{e}_{i_1} \otimes \mathbf{e}_{i_2} \otimes \ldots \otimes \mathbf{e}_{i_q})(\mathbf{e}_{j_1} \otimes \mathbf{e}_{j_2} \otimes \ldots \otimes \mathbf{e}_{j_q}) = \delta_{i_1 j_1} \ldots \delta_{i_q j_q},$$

and the right-hand side is zero if the set $(i_1, i_2, \ldots, i_q)$ differs from the set $(j_1, j_2, \ldots, j_q)$ and is equal to unity otherwise. Hence the basis $(\mathbf{e}_{i_1} \otimes \mathbf{e}_{i_2} \otimes \ldots \otimes \mathbf{e}_{i_q})$ of $E_n^{(q)}$, derived from the tensor product of an orthonormal basis of E_n, is itself orthonormal.

The covariant components of the tensor **T**, considered as a vector of the Euclidean space $E_n^{(q)}$, are clearly identical with the covariant components of **T** considered as a tensor in E_n.

IV. OUTER PRODUCTS

46. Antisymmetric tensors of order 2. Let us return to the case where E_n is a *general vector space* which we refer to an arbitrary basis $(\mathbf{e}_i)$. In the space $E_n^{(2)}$ referred to the associated basis $\mathbf{e}_i \otimes \mathbf{e}_j$ consider the antisymmetric tensors

$$\mathbf{T} = t^{ij} \mathbf{e}_i \otimes \mathbf{e}_j = \sum_{i<j} t^{ij} \mathbf{e}_i \otimes \mathbf{e}_j + \sum_{i \geqslant j} t^{ij} \mathbf{e}_i \otimes \mathbf{e}_j \quad (46.1)$$

where $\qquad t^{ij} = -t^{ji}, \qquad t^{ii} = 0.$

Interchanging the indices on the right-hand side and taking the antisymmetry into account it follows that

$$\mathbf{T} = \sum_{i<j} t^{ij} (\mathbf{e}_i \otimes \mathbf{e}_j - \mathbf{e}_j \otimes \mathbf{e}_i). \quad (46.2)$$

Any antisymmetric tensor of $E_n^{(2)}$ can therefore be expressed as a linear combination of the $^nC_2$ elements

$$(\mathbf{e}_i \otimes \mathbf{e}_j - \mathbf{e}_j \otimes \mathbf{e}_i), \qquad (i < j) \tag{46.3}$$

and these $^nC_2$ elements obviously form a linearly independent system, otherwise the $(\mathbf{e}_i \otimes \mathbf{e}_j)$ could not be linearly independent. This can be formulated as follows.

THEOREM: *The antisymmetric tensors in $E_n^{(2)}$ generate a vector sub-space $\Lambda_n^{(2)}$ of $E_n^{(2)}$ having $^nC_2$ dimensions. This sub-space has the elements of order 2 defined by (46.3) as a basis.*

47. Outer product of two vectors. DEFINITION: *Given two vectors $\mathbf{x}$ and $\mathbf{y}$ of E_n, the antisymmetric tensor*

$$\mathbf{x} \wedge \mathbf{y} = \mathbf{x} \otimes \mathbf{y} - \mathbf{y} \otimes \mathbf{x} \tag{47.1}$$

is called the outer product of the two vectors.

If x^i and y^i denote the components of $\mathbf{x}$ and $\mathbf{y}$ in $(\mathbf{e}_i)$, $\mathbf{x} \wedge \mathbf{y}$ has the antisymmetric components

$$P^{ij} = x^i y^j - x^j y^i \tag{47.2}$$

in $(\mathbf{e}_i \otimes \mathbf{e}_j)$.

The outer product of two vectors has the following properties, which are also sufficient to define it:

(*a*) If $\mathbf{x}, \mathbf{y}, \mathbf{z}$ denote vectors of E_n and α is a scalar, the outer product has the usual distributive properties:

$$\mathbf{x} \wedge (\mathbf{y} + \mathbf{z}) = \mathbf{x} \wedge \mathbf{y} + \mathbf{x} \wedge \mathbf{z},$$
$$(\mathbf{x} + \mathbf{y}) \wedge \mathbf{z} = \mathbf{x} \wedge \mathbf{z} + \mathbf{y} \wedge \mathbf{z},$$
$$\alpha \mathbf{x} \wedge \mathbf{y} = \mathbf{x} \wedge \alpha \mathbf{y} = \alpha(\mathbf{x} \wedge \mathbf{y}).$$

(*b*) It is anticommutative:

$$\mathbf{x} \wedge \mathbf{y} = -\mathbf{y} \wedge \mathbf{x},$$

in particular

$$\mathbf{x} \wedge \mathbf{x} = 0.$$

(*c*) If $(\mathbf{e}_1, \mathbf{e}_2, \ldots, \mathbf{e}_n)$ is a basis of E_n the $^nC_2$ elements

$$\mathbf{e}_i \wedge \mathbf{e}_j \qquad (i < j) \tag{47.3}$$

form a basis for $\Lambda_n^{(2)}$.

An outer product of two vectors is called a *bivector*.

48. Proper components of a bivector. Change of basis. According to (46.2) the antisymmetric tensor **T** may be written

$$\mathbf{T} = \sum_{i<j} t^{ij} \mathbf{e}_i \wedge \mathbf{e}_j.$$

Considered as an element of the space $\Lambda_n^{(2)}$ and referred to the basis (47.3), this tensor has the $^nC_2$ components $t^{ij}\,(i<j)$. These components are called the *proper components* of **T**. In order to distinguish them from the n^2 components of **T** considered as an element of $E_n^{(2)}$ they are represented by $t^{(ij)}$, where it is always understood that the index i is less than the index j.

Let us determine the transformation properties of the proper components on passing from the basis $(\mathbf{e}_i)$ to another basis $(\mathbf{e}_{i'})$ in E_n. The spaces $E_n^{(2)}$ and $\Lambda_n^{(2)}$ will be referred to the bases $(\mathbf{e}_{i'} \otimes \mathbf{e}_{j'})$ and $(\mathbf{e}_{i'} \wedge \mathbf{e}_{j'})$ respectively and, with the usual notation, we have

$$t^{ij} = A^i_{k'} A^j_{l'} t^{k'l'}.$$

Therefore

$$t^{(ij)} = \sum_{k'<l'} A^i_{k'} A^j_{l'} t^{k'l'} + \sum_{k'>l'} A^i_{k'} A^j_{l'} t^{k'l'}.$$

Interchanging the indices $k'l'$ in the second term and allowing for the antisymmetry of the transformed components $t^{k'l'}$ it follows that

$$t^{(ij)} = \sum_{k'<l'} (A^i_{k'} A^j_{l'} - A^i_{l'} A^j_{k'}) t^{(k'l')}.$$

Thus the proper components of **T** transform under a change of basis in E_n according to

$$t^{(ij)} = \sum_{k'<l'} \begin{vmatrix} A^i_{k'} & A^j_{k'} \\ A^i_{l'} & A^j_{l'} \end{vmatrix} t^{(k'l')}. \tag{48.1}$$

49. Outer forms of order 2. We have seen that each vector space E_n can be associated with a dual vector space, E_n^*, of linear forms defined on E_n. Outer products of order 2 can also be defined in E_n^*; they are the affine antisymmetric covariant tensors of order 2. These tensors define a vector sub-space $\Lambda_n^{*(2)}$ of $E_n^{*(2)}$, again having $^nC_2$ dimensions.

DEFINITION: *An outer form of order 2 is any element of $\Lambda_n^{*(2)}$.*

If E_n is referred to a basis $(\mathbf{e}_i)$ then we refer E_n^* to the dual basis (x^i)

and the spaces $E_n^{*(2)}$ and $\Lambda_n^{*(2)}$ to the bases $(x^i \otimes x^j)$ and $(x^i \wedge x^j)$. An element F of $\Lambda_n^{*(2)}$ which has the components f_{ij} in $E_n^{*(2)}$ can be written

$$F = \sum_{i<j} f_{(ij)} x^i \wedge x^j,$$

where $\qquad\qquad f_{(ij)} = f_{ij} \qquad$ for $i < j$.

The $^nC_2$ quantities $f_{(ij)}$ are the proper components of the form F. Under a change of basis in E_n they transform according to the formulae

$$f_{(ij)} = \sum_{i<j} \begin{vmatrix} A_i^{k'} & A_j^{k'} \\ A_i^{l'} & A_j^{l'} \end{vmatrix} f_{(k'l')}. \tag{49.1}$$

The scalar quantity

$$\sum_{i<j} f_{(ij)} t^{(ij)} = \tfrac{1}{2} \sum_{ij} f_{ij} t^{ij} \tag{49.2}$$

is said to be the *value of the form F* for an element $\mathbf{T}$ of $\Lambda_n^{(2)}$ having proper components $t^{(ij)}$.

If the tensor $\mathbf{T}$ varies over $\Lambda_n^{(2)}$ the quantity (49.2) constitutes a linear form defined on $\Lambda_n^{(2)}$. Conversely, given a linear form defined on $\Lambda_n^{(2)}$, this can always be written

$$\sum_{i<j} f_{(ij)} t^{(ij)}. \tag{49.3}$$

The $t^{(ij)}$ transform under a change of basis according to (48.1) and, since (49.3) is a scalar, the $f_{(ij)}$ transform according to (49.1) and are therefore the proper components of an element of $\Lambda_n^{*(2)}$. So the affine antisymmetric covariant tensors of order 2 can be interpreted equally well as elements of $\Lambda_n^{*(2)}$ or elements of the space $[\Lambda_n^{(2)}]^*$ dual to $\Lambda_n^{(2)}$.

50. Completely antisymmetric tensors. In view of their importance we have restricted ourselves so far to the study of antisymmetric tensors of order 2. Analogous results can be shown to hold for completely antisymmetric tensors of order $q \leqslant n$, that is for tensors which are antisymmetric with respect to every pair of indices. Contravariant tensors of this type constitute a vector sub-space of $E_n^{(q)}$ in ^nC_q dimensions. ^nC_q proper components are therefore sufficient to define them.

We begin by indicating the results for $q = n$. A completely antisymmetric tensor of order n has only one proper component $t^{(12\ldots n)}$. Because of the antisymmetry the ordinary components of this tensor are given by

$$t^{i_1 i_2 \ldots i_n} = \varepsilon^{i_1 i_2 \ldots i_n} t^{(12\ldots n)}, \tag{50.1}$$

where $\varepsilon^{i_1 i_2 \ldots i_n}$ is equal to zero if any two indices are equal, to $+1$ if the permutation $(i_1, i_2, \ldots, i_3)$ of the set $(1, 2, \ldots, n)$ is even and to -1 otherwise.

Under a change of the basis $(\mathbf{e}_i)$ the proper component $t^{(12\ldots n)}$ transforms according to the formula

$$t^{(12\ldots n)} = \Delta t^{(1'2'\ldots n')} \tag{50.2}$$

where Δ is the determinant

$$\Delta = \begin{vmatrix} A_{1'}^1 & A_{1'}^2 & \ldots & A_{1'}^n \\ A_{2'}^1 & A_{2'}^2 & \ldots & \ldots \\ \vdots & \ldots & \ldots & \ldots \\ A_{n'}^1 & \ldots & \ldots & A_{n'}^n \end{vmatrix}. \tag{50.3}$$

51. Outer products in a Euclidean space. Suppose E_n is a Euclidean space. Then the various tensors defined by outer products of vectors are Euclidean. The contravariant and covariant components of a Euclidean tensor of order 2 are related by

$$t_{ij} = g_{ik} g_{jl} t^{kl}.$$

If the tensor under consideration is antisymmetric, its contravariant and covariant proper components are related by

$$t_{(ij)} = \sum_{k<l} \begin{vmatrix} g_{ik} & g_{jk} \\ g_{il} & g_{jl} \end{vmatrix} t^{(kl)} \tag{51.1}$$

as can readily be established by an argument similar to that given in §46 and §48. The space $\Lambda_n^{(2)}$ has the structure of a Euclidean space, its fundamental quadratic form having the coefficients $(g_{ik}g_{jl} - g_{il}g_{jk})$.

For a completely antisymmetric tensor of order n, similar reasoning shows that

$$t_{(12\ldots n)} = g t^{(12\ldots n)}, \tag{51.2}$$

where g designates the determinant constructed from the g_{ij}.

52. Adjoint tensor of a completely antisymmetric tensor. When E_n is a Euclidean space there is a particularly interesting completely antisymmetric tensor of order n which is directly associated with the determinant g. For simplicity we assume the space to be properly Euclidean.†

It is of interest to find a geometrical interpretation of g in the Euclidean vector space of elementary geometry. If $\mathbf{x}$, $\mathbf{y}$, $\mathbf{z}$ denote three arbitrary vectors with components x^i, y^i, z^i referred to an orthonormal basis in this space, then the volume V of the parallelopiped, which has the three vectors $\mathbf{x}$, $\mathbf{y}$, $\mathbf{z}$ as edges, is equal to the absolute value of the determinant

$$\begin{vmatrix} x^1 & x^2 & x^3 \\ y^1 & y^2 & y^3 \\ z^1 & z^2 & z^3 \end{vmatrix}.$$

Its square

$$V^2 = \begin{vmatrix} \sum (x^i)^2 & \sum x^i y^i & \sum x^i z^i \\ \sum x^i y^i & \sum (y^i)^2 & \sum y^i z^i \\ \sum x^i z^i & \sum y^i z^i & \sum (z^i)^2 \end{vmatrix}$$

can be written, in terms of invariants, as

$$V^2 = \begin{vmatrix} \mathbf{x}^2 & \mathbf{x} \cdot \mathbf{y} & \mathbf{x} \cdot \mathbf{z} \\ \mathbf{x} \cdot \mathbf{y} & \mathbf{y}^2 & \mathbf{y} \cdot \mathbf{z} \\ \mathbf{x} \cdot \mathbf{z} & \mathbf{y} \cdot \mathbf{z} & \mathbf{z}^2 \end{vmatrix}. \tag{52.1}$$

Now let $(\mathbf{e}_1, \mathbf{e}_2, \mathbf{e}_3)$ be any basis of the space under consideration. The elements of the determinant in (52.1) are just the g_{ij} which correspond to this basis if $\mathbf{e}_1$, $\mathbf{e}_2$, $\mathbf{e}_3$ are taken to be the vectors $\mathbf{x}$, $\mathbf{y}$, $\mathbf{z}$. Hence, if V is the volume of the parallelopiped constructed on $(\mathbf{e}_1, \mathbf{e}_2, \mathbf{e}_3)$, we have

$$V^2 = g.$$

Having established this result let us return to the proper Euclidean space P_n and make the change of basis which carries $(\mathbf{e}_i)$ into $(\mathbf{e}_{j'})$.

† This ensures that the determinant g is positive and therefore simplifies manipulations involving $\sqrt{g}$. (T.)

How does the determinant g transform under this change of basis? We know that its elements transform according to the tensor rule

$$g_{k'l'} = A^i_{k'} A^j_{l'} g_{ij}.$$

It then follows from a standard theorem on the multiplication of determinants that, if g' denotes the determinant formed from the $g_{k'l'}$,

$$g' = \Delta^2 g$$

where Δ denotes the determinant (50.3) constructed from the $A^i_{k'}$. Taking the square root of both sides it follows that

$$\frac{1}{\sqrt{g}} = |\Delta| \frac{1}{\sqrt{g'}}. \tag{52.2}$$

Let us limit further developments to changes of basis such that the determinant Δ is positive. We say that this restricts us to bases which have the same *sense* or *handedness* as (e_i). Under such changes of basis, the quantity $1/\sqrt{g}$ transforms like the proper contravariant component of a completely antisymmetric tensor of order n (see (50.2)); according to (51.2) this tensor has the proper covariant component $\sqrt{g}$. Its ordinary contravariant components may be written

$$\epsilon^{i_1 i_2 \ldots i_n} = \varepsilon^{i_1 i_2 \ldots i_n} \frac{1}{\sqrt{g}}. \tag{52.3}$$

while its covariant components are given by

$$\epsilon_{i_1 i_2 \ldots i_n} = \varepsilon_{i_1 i_2 \ldots i_n} \sqrt{g}, \tag{52.4}$$

where the quantities $\varepsilon_{i_1 i_2 \ldots i_n}$ are numerically equal to the $\varepsilon^{i_1 i_2 \ldots i_n}$.

Given a completely antisymmetric tensor $\mathbf{T}$ of order $q(\leqslant n)$ with components $t^{i_1 i_2 \ldots i_q}$ or $t_{i_1 i_2 \ldots i_q}$ the completely antisymmetric tensor $\mathbf{T}'$ of order $(n-q)$ obtained by contracted multiplication of $\mathbf{T}$ with (52.3) is called the *adjoint tensor* of $\mathbf{T}$. Its components are given by

$$t'^{i_{q+1} \ldots i_n} = \frac{1}{q!} \epsilon^{i_1 \ldots i_q i_{q+1} \ldots i_n} t_{i_1 \ldots i_q}, \tag{52.5}$$

$$t'_{i_{q+1} \ldots i_n} = \frac{1}{q!} \epsilon_{i_1 \ldots i_q i_{q+1} \ldots i_n} t^{i_1 \ldots i_q}. \tag{52.6}$$

The completely antisymmetric tensors of orders n and $(n-1)$ have adjoint tensors which are respectively a scalar and a vector.

Consider, as an example, the Euclidean space of elementary geometry, which we assume to be referred to an orthonormal basis so that contravariant and covariant components are identical. Let $\mathbf{x} \wedge \mathbf{y}$ be a bivector of this space and consider its adjoint $\mathbf{z}$. Denoting the components of $\mathbf{x}$, $\mathbf{y}$, $\mathbf{z}$ by x^i, y^i, z^i, the components of the bivector $\mathbf{x} \wedge \mathbf{y}$ are given by

$$\left. \begin{aligned} P^{23} &= x^2 y^3 - x^3 y^2, \\ P^{31} &= x^3 y^1 - x^1 y^3, \\ P^{12} &= x^1 y^2 - x^2 y^1, \end{aligned} \right\}$$

and if the determinant g is equal to unity (52.5) gives

$$\left. \begin{aligned} z^1 &= x^2 y^3 - x^3 y^2, \\ z^2 &= x^3 y^1 - x^1 y^3, \\ z^3 &= x^1 y^2 - x^2 y^1. \end{aligned} \right\} \tag{52.7}$$

In elementary vector analysis, the vector $\mathbf{z}$ is known as the *vector product* of the two vectors $\mathbf{x}$ and $\mathbf{y}$. The existence of such a vector product is clearly related to the fact that the space of elementary geometry is three dimensional.

CHAPTER IV

Curvilinear Coordinates in Euclidean Space

I. DERIVATIVES AND DIFFERENTIALS OF VECTORS AND POINTS

53. Vector derivative of a vector. Let E_n be an n-dimensional vector space. If a vector $\mathbf{x}$ in E_n corresponds to each value t of a scalar variable in the interval (a, b), we say that $\mathbf{x}$ is a function of t and write $\mathbf{x} = \mathbf{x}(t)$.

We now introduce a positive definite quadratic form so that E_n is a proper Euclidean space. The variable vector $\mathbf{x}$ is then said to *tend to the zero vector* (or more briefly, to *tend to zero*) if the scalar $|\mathbf{x}|$ tends to zero. It is clear that this definition is independent of the particular quadratic form chosen.

We say that the vector $\mathbf{x}(t)$ is a *continuous function* of t if, when t is increased by $\varDelta t$, the vector

$$\varDelta \mathbf{x} = \mathbf{x}(t + \varDelta t) - \mathbf{x}(t)$$

tends to zero as $\varDelta t \to 0$. If there exists a vector $\mathbf{x}'$ such that

$$\frac{\varDelta \mathbf{x}}{\varDelta t} - \mathbf{x}'$$

tends to zero as $\varDelta t \to 0$, $\mathbf{x}'$ is said to be the *vector derivative* of $\mathbf{x}$ for the value t of the variable. The vector

$$d\mathbf{x} = \mathbf{x}' \, dt$$

is called the *differential* of $\mathbf{x}$ and we write $\mathbf{x}'$ as $d\mathbf{x}/dt$.

These definitions are simple generalizations of the analogous definitions of elementary vector calculus. The formulae for the derivatives of a sum, of a product of a vector and a scalar, and of a vector function of a scalar function are formally identical with the well-known formulae of elementary vector calculus and are demonstrated in exactly the same way. This also holds for the derivative of the scalar product.

54. Vector derivative of a point. Let $\mathscr{E}_n$ be an n-dimensional affine point space. Consider a scalar t which varies over the interval (a, b). If to each value of t there corresponds a point M of $\mathscr{E}_n$, M is said to be a *function* of t and we write $M = M(t)$.

If O is an arbitrary fixed point of $\mathscr{E}_n$ the vector $\mathbf{x} = \overrightarrow{OM}$ is a function of t. Suppose that this vector has the derivative $\mathbf{x}'$. It is obvious that the vector $\mathbf{x}'$ does not depend on the choice of the fixed point O, but only on M. In fact, if O' denotes another arbitrary fixed point,

$$\overrightarrow{OM} = \overrightarrow{OO'} + \overrightarrow{O'M}$$

and, as $\overrightarrow{OO'}$ is fixed,

$$\frac{d}{dt}(\overrightarrow{OM}) = \frac{d}{dt}(\overrightarrow{O'M}) = \mathbf{x}'.$$

The vector $\mathbf{x}'$ is termed the *vector derivative* of the point M and it is represented by $\mathbf{M}'$. The vector

$$d\mathbf{M} = \mathbf{M}' \, dt$$

is called the *differential* of M and we write $\mathbf{M}' = d\mathbf{M}/dt$.

55. Vector functions of several scalar variables. As in elementary vector analysis, a vector $\mathbf{x}$ can be a function of a number of independent scalar variables α, β, γ. The concept of a *partial derivative* can immediately be extended to such a vector function. As in ordinary analysis the relation

$$\frac{\partial^2 \mathbf{x}}{\partial \alpha \, \partial \beta} = \frac{\partial^2 \mathbf{x}}{\partial \beta \, \partial \alpha}$$

may be established provided that the partial derivatives are continuous. The differential of such a vector function is given by

$$d\mathbf{x} = \frac{\partial \mathbf{x}}{\partial \alpha} d\alpha + \frac{\partial \mathbf{x}}{\partial \beta} d\beta + \frac{\partial \mathbf{x}}{\partial \gamma} d\gamma.$$

If $\mathbf{x}$ is a function of t through several scalar variables $\alpha(t)$, $\beta(t)$, $\gamma(t)$ its derivative is obtained in the same way as in ordinary analysis:

$$\frac{d\mathbf{x}}{dt} = \frac{\partial \mathbf{x}}{\partial \alpha} \frac{d\alpha}{dt} + \frac{\partial \mathbf{x}}{\partial \beta} \frac{d\beta}{dt} + \frac{\partial \mathbf{x}}{\partial \gamma} \frac{d\gamma}{dt}.$$

The above remarks concerning the vector derivatives of a vector apply equally well to the vector derivatives of a point.

II. CURVILINEAR COORDINATES IN A EUCLIDEAN POINT SPACE

56. Curvilinear coordinates. Associated reference frames. We shall restrict the remainder of this chapter to n-dimensional *Euclidean point spaces* $\mathscr{E}_n$ and we propose to study several geometrical ideas. Let $\mathscr{E}_n$ be referred at some moment to an arbitrary frame and denote the coordinates of any point M with respect to this frame by (x^i). We have already seen that there corresponds one, and only one, point M of $\mathscr{E}_n$ to each system of n numbers $(x^1, x^2, \ldots, x^n)$, and conversely. In order to distinguish these coordinates from those which are introduced below we shall call them *rectilinear coordinates*.†

Consider n continuously differentiable functions $f^i(y^1, y^2, \ldots, y^n)$ of n variables y^i and put

$$x^i = f^i(y^1, y^2, \ldots, y^n) \quad (i = 1, 2, \ldots, n). \tag{56.1}$$

These n functions are assumed to be independent so that, when the variables (y^i) vary over a domain $\mathscr{D}'$, the system of n equations (56.1) can be solved for the y^i, giving

$$y^i = g^i(x^1, x^2, \ldots, x^n) \quad (i = 1, 2, \ldots, n). \tag{56.2}$$

† The essential distinction between a rectilinear coordinate system and a curvilinear one is that in the latter the reference frame $(\mathbf{e}_i)$ is a function of position while in the former it is constant. (T.)

Here the point M, whose coordinates are (x^i), varies over a certain domain $\mathcal{D}$ of $\mathcal{E}_n$. Since the n functions (56.1) are assumed to be independent, the *functional determinant* or *Jacobian*

$$\frac{D(x^1, x^2, \ldots, x^n)}{D(y^1, y^2, \ldots, y^n)} = \begin{vmatrix} \dfrac{\partial f^1}{\partial y^1} & \cdots & \dfrac{\partial f^n}{\partial y^1} \\ \vdots & & \\ \dfrac{\partial f^1}{\partial y^n} & \cdots & \dfrac{\partial f^n}{\partial y^n} \end{vmatrix} \tag{56.3}$$

is different from zero in $\mathcal{D}'$. This is also true of the functional determinant

$$\frac{D(y^1 y^2 \ldots y^n)}{D(x^1 x^2 \ldots x^n)}$$

which is the inverse of (56.3).

This demonstrates the existence of a one-to-one correspondence between the points M of $\mathcal{D}$ and the variables (y^i) which range over $\mathcal{D}'$. If the functions f^i are non-linear the y^i cannot be interpreted as a system of rectilinear coordinates. The point M is a function, many times continuously differentiable, of the n scalar variables (y^i). We say that the space $\mathcal{E}_n$ has been referred, in the domain $\mathcal{D}$, to a system of *curvilinear coordinates* (y^i). The curves traced out by a point M for which only one of the (y^i) varies are called *coordinate lines*; n of these intersect at any point M of $\mathcal{E}_n$. In rectilinear coordinate systems the coordinate lines are straight which accounts for the name given to these coordinates.

Given a curvilinear coordinate system (y^i) and a point M of $\mathcal{E}_n$ we associate with M a reference frame called the *natural frame* at M of the system (y^i). This reference frame has its origin at the point M and its vectors are given by

$$\mathbf{e}_i = \frac{\partial \mathbf{M}}{\partial y^i} \quad (i = 1, 2, \ldots, n). \tag{56.4}$$

It is obvious that the system of n vectors $\mathbf{e}_i$ is linearly independent since the determinant of the components of these vectors referred to the initial reference frame is the determinant (56.3) which is non-zero. The n vectors $\mathbf{e}_i$ are manifestly collinear with the tangents to the n coordinate lines which intersect at M. From (56.4) the vector differential of M is given by

$$d\mathbf{M} = \mathbf{e}_i dy^i. \qquad (56.5)$$

In other words the n quantities dy^i are the contravariant components of the vector $d\mathbf{M}$ in the natural reference frame of the system (y^i) at M.

We speak of transforming the curvilinear coordinates if we substitute a new system of variables $(y^{j'})$ for the (y^i). We then have

$$y^{j'} = y^{j'}(y^1, y^2, \ldots, y^n), \qquad y^i = y^i(y^{1'}, y^{2'}, \ldots, y^{n'}), \qquad (56.6)$$

where the $y^{j'}$ are many times continuously differentiable functions of the y^i, and conversely. When such a transformation of curvilinear coordinates is made the natural reference system $(M, \mathbf{e}_i)$ of the system (y^i) is replaced by the natural reference system $(M, \mathbf{e}_{j'})$ of the system $(y^{j'})$. It is easy to deduce the transformation rules which link one reference system to another. From the formula for the derivative of a function of a function we have, in fact

$$\mathbf{e}_{j'} = \frac{\partial \mathbf{M}}{\partial y^{j'}} = \frac{\partial \mathbf{M}}{\partial y^i} \frac{\partial y^i}{\partial y^{j'}}$$

It follows that

$$\mathbf{e}_{j'} = \frac{\partial y^i}{\partial y^{j'}} \mathbf{e}_i;$$

and conversely

$$\mathbf{e}_i = \frac{\partial y^{j'}}{\partial y^i} \mathbf{e}_{j'}.$$

The basis $(\mathbf{e}_i)$ of the Euclidean vector space associated with $\mathscr{E}_n$ has thus been replaced by the basis $(\mathbf{e}_{j'})$ given by the linear transformation

$$\mathbf{e}_i = A_i^{j'} \mathbf{e}_{j'}; \qquad \mathbf{e}_{j'} = A_{j'}^i \mathbf{e}_i; \qquad (56.7)$$

where

$$A_i^{j'} = \frac{\partial y^{j'}}{\partial y^i}; \qquad A_{j'}^i = \frac{\partial y^i}{\partial y^{j'}}. \qquad (56.8)$$

This may be stated formally as follows:

THEOREM: *Each change of curvilinear coordinates is associated with the change of natural reference frame at M defined by the relations (56.7) and (56.8).*

57. Example of a curvilinear coordinate system. Consider the Euclidean space, $\mathscr{E}_3$, of elementary geometry and let $Oxyz$ be a Cartesian reference frame. The cylindrical coordinate system and the polar coordinate system in this space are examples of curvilinear coordinates. The latter system, for example, is defined by

$$x = r\sin\theta\cos\psi, \quad y = r\sin\theta\sin\psi, \quad z = r\cos\theta,$$

and, conversely

$$r = \sqrt{(x^2+y^2+z^2)}, \quad \psi = \tan^{-1}y/x, \quad \theta = \tan^{-1}\frac{\sqrt{(x^2+y^2)}}{z}.$$

Let us write

$$y^1 = r, \quad y^2 = \psi, \quad y^3 = \theta.$$

The coordinate lines at M are respectively the radius vector $\overrightarrow{OM}$, the circle of radius $r\sin\theta$ through M with its centre on and plane perpendicular to the axis Oz and the meridian circle, centre O, radius r, which passes through M. The natural reference frame at M consists of $\mathbf{e}_1$, the unit vector in the direction OM, $\mathbf{e}_2$ tangential to the circle about Oz and of length $r\sin\theta$, and $\mathbf{e}_3$ the vector tangent to the meridian and of length r. We note that these three vectors are always orthogonal.

In general, when the n vectors $\mathbf{e}_i$ of the natural reference system are orthogonal, the curvilinear coordinate system is itself said to be orthogonal.

58. The line element. If the Euclidean point space $\mathscr{E}_n$ is referred to a system of curvilinear coordinates (y^i), the vector $d\mathbf{M}$ has components dy^i with respect to the natural reference frame at M. The square of this vector, or the square of the distance between two neighbouring points, is therefore given by

$$ds^2 = g_{ij}dy^i dy^j, \tag{58.1}$$

where

$$g_{ij} = \mathbf{e}_i \cdot \mathbf{e}_j.$$

As the point M varies, the natural reference frame also varies. The scalar products g_{ij} of two vectors of the reference frame are therefore functions of the curvilinear coordinates of the point M. This is the case, for instance, in the example of the previous section, where

$$ds^2 = dr^2 + r^2\cos^2\theta\, d\psi^2 + r^2 d\theta^2. \tag{58.2}$$

The expression (58.1) is called the *line element* or *metric*. The length of arc of any curve in $\mathscr{E}_n$ can be calculated from a knowledge of the metric. If the curve $\widehat{AB}$ is defined by giving the y^i as functions of the parameter t which varies over the interval (a,b) its length is given by the integral

$$\widehat{AB} = \int_a^b \sqrt{\left(g_{ij} \frac{dy^i}{dt} \frac{dy^j}{dt} \right)} \, dt. \tag{58.3}$$

In the same way, the element of volume in this space is the volume of a parallelopiped constructed from the vectors $\mathbf{e}_1 dy^1, \mathbf{e}_2 dy^2, \ldots, \mathbf{e}_n dy^n$; i.e.

$$dV = \sqrt{(|g|)} \, dy^1 \, dy^2 \, dy^3 \ldots dy^n, \tag{58.4}$$

where g denotes the determinant of the g_{ij}. Finite volumes can be determined from this by integration.

59. Tensor fields. A Euclidean vector space E_n is associated with the Euclidean point space $\mathscr{E}_n$. Each reference frame of $\mathscr{E}_n$ defines a basis in E_n and, consequently, bases for the various tensor powers of E_n. For brevity we shall say that the components of a Euclidean tensor with respect to such a basis are the components referred to the corresponding frame of $\mathscr{E}_n$.

Suppose that we associate with each point M of $\mathscr{E}_n$ a Euclidean tensor defined by its components referred to the natural frame at M of the system (y^i). We call this a *tensor field* in the system of curvilinear coordinates (y^i). The remainder of this chapter is devoted to tensor analysis, which is essentially the study of tensor fields.

We have seen that, to each change of curvilinear coordinates there corresponds a change of the natural reference frame at M defined by (56.7), (56.8). If **T**, for example, denotes a tensor of order three with the mixed components $t^{ij}{}_k$, these components transform under a change of curvilinear coordinates according to the tensor relations

$$t^{ij}{}_k = A^i_{l'} A^j_{m'} A^{n'}_k t^{l'm'}{}_{n'},$$

where the A's are defined by (56.8).

The quantities g_{ij} are the covariant components of a tensor referred to the natural frame at M. The fundamental tensor is therefore an example of a tensor field and its components g_{ij} transform according to

$$g_{ij} = A_i^{k'} A_j^{l'} g_{k'l'}.$$

III. THE CHRISTOFFEL SYMBOLS

60. The fundamental problem of tensor analysis. Let the Euclidean space $\mathscr{E}_n$ be referred to a curvilinear coordinate system (y^i) and let its metric be

$$ds^2 = g_{ij} dy^i dy^j. \tag{60.1}$$

In the study of tensor fields referred to curvilinear coordinate systems the following difficulty is encountered: tensors associated with different points are referred to different reference frames and, in order to compare tensors at neighbouring points with one another, it is necessary to consider the variation of the reference frame between those points.

Since

$$g_{ij} = \mathbf{e}_i \cdot \mathbf{e}_j,$$

if the metric (60.1) is a known function of position then we know the form of the natural reference frame $(\mathbf{e}_i)$ for the different points of $\mathscr{E}_n$. This leads us to pose the following problem:

Given the Euclidean space $\mathscr{E}_n$ referred to a curvilinear coordinate system (y^i) and with the metric

$$ds^2 = g_{ij} dy^i dy^j,$$

determine the natural reference frame $(M + d\mathbf{M}, \mathbf{e}_i + d\mathbf{e}_i)$† at the point $M + d\mathbf{M}$ in the immediate neighbourhood of M in terms of $(M, \mathbf{e}_i)$.

The natural frame $(M + d\mathbf{M}, \mathbf{e}_i + d\mathbf{e}_i)$ will be completely determined relative to $(M, \mathbf{e}_i)$ if we know the contravariant components of the vectors $d\mathbf{M}$ and $d\mathbf{e}_i$ referred to the basis $(\mathbf{e}_i)$. The contravariant components, dy^i, of $d\mathbf{M}$ are given by

$$d\mathbf{M} = dy^i \mathbf{e}_i. \tag{60.2}$$

† It is convenient to use the purely symbolic notation $M + d\mathbf{M}$ to represent a point displaced by $d\mathbf{M}$ from the point M. (T.)

We can also write

$$de_i = \omega^j{}_i e_j, \tag{60.3}$$

where the $\omega^j{}_i$ denote the contravariant components† of the de_i. These components are obviously linearly dependent on the dy^k. We therefore have

$$\omega^j{}_i = \Gamma_k{}^j{}_i dy^k \tag{60.4}$$

where the $\Gamma_k{}^j{}_i$ denote n^3 functions of the variables (y^k). Our problem is therefore reduced to the determination of the n^3 functions $\Gamma_k{}^j{}_i$ from the $[n(n+1)]/2$ functions g_{ij}.

61. Relations between the $\Gamma_k{}^j{}_i$. (*I*) Given the line element (60.1), the natural reference frame at every point M of $\mathscr{E}_n$ satisfies

$$e_i \cdot e_j = g_{ij}. \tag{61.1}$$

It follows that

$$e_i \cdot de_j + e_j \cdot de_i = dg_{ij}.$$

Substituting for the de_i from (60.3) and using (61.1) we obtain

$$g_{ik} \omega^k{}_j + g_{jk} \omega^k{}_i = dg_{ij}. \tag{61.2}$$

The form of (61.2) leads us to introduce the covariant components† ω_{ji} of de_i in addition to the quantities $\omega^j{}_i$ and the coefficients Γ_{kji} of dy^k in the expression for ω_{ji}. We then have

$$\omega_{ji} = \Gamma_{kji} dy^k,$$

where

$$\omega_{ji} = g_{jk} \omega^k{}_i; \quad \Gamma_{kji} = g_{jh} \Gamma_k{}^h{}_i; \quad \Gamma_k{}^j{}_i = g^{jh} \Gamma_{khi}. \tag{61.3}$$

A knowledge of the n^3 functions Γ_{kji} is thus equivalent to knowing the n^3 functions $\Gamma_k{}^j{}_i$. Using this notation (61.2) can be rewritten as

$$\omega_{ij} + \omega_{ji} = dg_{ij}. \tag{61.4}$$

† This terminology should not be taken to imply that the ω^j are the components of a tensor (see § **63**). (T.)

Identifying the coefficients of dy^k on both sides of (61.4) we obtain an equivalent system of equations relating finite quantities:

$$\Gamma_{kij} + \Gamma_{kji} = \partial_k g_{ij}, \tag{61.5}$$

where we have written

$$\partial_k g_{ij} = \frac{\partial g_{ij}}{\partial y^k}$$

in order to emphasize the covariant nature of the suffix k.

The system (61.5) comprises as many equations as there are distinct quantities $\partial_k g_{ij}$. Since there are $[n(n+1)]/2$ quantities g_{ij} (61.5) contains $[n^2(n+1)]/2$ equations.

(2) The equations (60.2) and (60.3) give the differentials of the point M and of the vectors $\mathbf{e}_i$ respectively and are therefore integrable and thus give second derivatives of M and of the $\mathbf{e}_i$ which are symmetrical with respect to the indices of differentiation. Let us determine these integrability conditions for equations (60.2); we have

$$\frac{\partial^2 \mathbf{M}}{\partial y^k \, \partial y^j} = \frac{\partial}{\partial y^k}\left(\frac{\partial \mathbf{M}}{\partial y^j}\right) = \frac{\partial \mathbf{e}_j}{\partial y^k}$$

and substituting (60.3),

$$\frac{\partial^2 \mathbf{M}}{\partial y^k \, \partial y^j} = \Gamma_k{}^h{}_j \mathbf{e}_h. \tag{61.6}$$

Similarly

$$\frac{\partial^2 \mathbf{M}}{\partial y^j \, \partial y^k} = \Gamma_j{}^h{}_k \mathbf{e}_h. \tag{61.7}$$

Since the left-hand sides of (61.6) and (61.7) are equal we deduce that

$$\Gamma_k{}^h{}_j \mathbf{e}_h = \Gamma_j{}^h{}_k \mathbf{e}_h$$

and hence

$$\Gamma_k{}^h{}_j = \Gamma_j{}^h{}_k, \tag{61.8}$$

which may, with the aid of (61.3), be written as

$$\Gamma_{kij} = \Gamma_{jik}. \tag{61.9}$$

The quantities $\Gamma_k{}^j{}_i$ are therefore symmetric with respect to their lower indices and the quantities Γ_{kji} with respect to their first and last indices. For each value of i the integrability conditions (61.9) give $[n(n-1)]/2$ separate equations. The system (61.9) thus comprises a total of $[n^2(n-1)]/2$ equations. These equations added to the $[n^2(n+1)]/2$ equations (61.5) give a system of n^3 linear equations in the n^3 unknowns Γ_{kji}.

The conditions for the integrability of the de_i could be determined in a similar manner. We shall show, however, that the n^3 relations already obtained allow the calculation of the Γ_{kji} from the g_{ij} and their derivatives. It follows that the integrability conditions for the de_i are necessary criteria for the original problem to be soluble. This is associated with the fact that given an arbitrary quadratic form with variable coefficients there does not always exist a curvilinear coordinate system in the Euclidean space $\mathscr{E}_n$ which is consistent with the interpretation of this form as the ds^2 of that space. We shall return to this point later.

62. Explicit determination of the $\Gamma_k{}^j{}_i$. It is a simple matter to obtain an explicit solution of the n^3 linear equations (61.5) and (61.9). Allowing for (61.9), equations (61.5) may be written

$$\Gamma_{jik} + \Gamma_{kji} = \partial_k g_{ij}. \tag{62.1}$$

Cyclic permutation of the indices gives

$$\Gamma_{kji} + \Gamma_{ikj} = \partial_i g_{jk}, \tag{62.2}$$

$$\Gamma_{ikj} + \Gamma_{jik} = \partial_j g_{ki}. \tag{62.3}$$

Adding (62.1) and (62.2) and subtracting (62.3) we obtain

$$2\Gamma_{kji} = \partial_k g_{ij} + \partial_i g_{jk} - \partial_j g_{ki}.$$

Using the notation

$$[ki,j] = \tfrac{1}{2}(\partial_k g_{ij} + \partial_i g_{jk} - \partial_j g_{ki}), \tag{62.4}$$

this becomes

$$\Gamma_{kji} = [ki,j]. \tag{62.5}$$

It is clear that the values of Γ_{kji} given by (62.5) satisfy (61.5) and (61.9). Using (61.3) we obtain

$$\Gamma_k{}^j{}_i = g^{jh}\Gamma_{khi} = g^{jh}[ki, h].$$

This suggests another bracket symbol defined by

$$\{_k{}^j{}_i\} = g^{jh}[ki, h], \tag{62.6}$$

so that

$$\Gamma_k{}^j{}_i = \{_k{}^j{}_i\}. \tag{62.7}$$

The symbols defined by (62.4) and (62.6) are called *Christoffel symbols of the first and second kind* respectively. They provide a means of calculating the Γ_{kji} and $\Gamma_k{}^j{}_i$ from the g_{ij} and their derivatives. This completes the solution of our fundamental problem.

63. Transformation of the $\Gamma_k{}^j{}_i$. It is important to note that neither the $\omega^j{}_i$ nor the $\Gamma_k{}^j{}_i$ are the components of a tensor. Let us determine the way in which these quantities transform under a change of curvilinear coordinates. The vectors of the natural reference frame transform according to the relations

$$\mathbf{e}_i = A_i^{l'}\mathbf{e}_{l'},$$

and differentiating

$$d\mathbf{e}_i = A_i^{l'}d\mathbf{e}_{l'} + dA_i^{l'}\mathbf{e}_{l'}.$$

However, from (60.3),

$$d\mathbf{e}_i = \omega^j{}_i\mathbf{e}_j, \qquad d\mathbf{e}_{l'} = \omega^{m'}{}_{l'}\mathbf{e}_{m'},$$

giving

$$\omega^j{}_i\mathbf{e}_j = A_i^{l'}\omega^{m'}{}_{l'}\mathbf{e}_{m'} + dA_i^{l'}\mathbf{e}_{l'} = (A_i^{l'}A_{m'}^j\omega^{m'}{}_{l'} + A_{l'}^j dA_i^{l'})\mathbf{e}_j.$$

Equating the coefficients of $\mathbf{e}_j$ we see that the $\omega^j{}_i$ transform according to the formulae

$$\omega^j{}_i = A_i^{l'}A_{m'}^j\omega^{m'}{}_{l'} + A_{l'}^j dA_i^{l'}. \tag{63.1}$$

Expanding both sides of (63.1) in terms of the quantities $\Gamma_k{}^j{}_i$ we find that

$$\Gamma_k{}^j{}_i dy^k = A_i^{l'}A_{m'}^j\Gamma_{n'}{}^{m'}{}_{l'}dy^{n'} + A_{l'}^j\,\partial_k A_i^{l'}dy^k.$$

Since the dy^k are arbitrary displacements we can equate their co-efficients on both sides of this equation. Hence the $\Gamma_k{}^j{}_i$ transform according to the formula

$$\Gamma_k{}^j{}_i = A_i^{l'} A_{m'}^j A_k^{n'} \Gamma_{n'}{}^{m'}{}_{l'} + A_{l'}^j \, \partial_k A_i^{l'}, \qquad (63.2)$$

which involves the second order derivatives of the functions defining the transformation.

IV. ABSOLUTE DIFFERENTIALS AND COVARIANT DERIVATIVES

64. Absolute differential of a vector. (*1*) Let us consider a field of vectors **v** in $\mathscr{E}_n$ defined by their contravariant components v^i and try to determine the contravariant components of the differential $d\mathbf{v}$ of a vector of the field for an infinitesimal variation of the point M. On passing from M to $M + d\mathbf{M}$ not only do the components v^i change, but the natural frame is also modified in the way we have just seen. Since the vector **v** is defined at all points of $\mathscr{E}_n$ by the relation

$$\mathbf{v} = v^i \mathbf{e}_i,$$

we get by differentiation

$$d\mathbf{v} = dv^i \mathbf{e}_i + v^i d\mathbf{e}_i.$$

Substituting for $d\mathbf{e}_i$ from (60.3), and relabelling the indices,

$$d\mathbf{v} = dv^i \mathbf{e}_i + v^k \omega^i{}_k \mathbf{e}_i.$$

It follows that the contravariant components of the vector $d\mathbf{v}$ are given by

$$\nabla v^i = dv^i + \omega^i{}_k v^k. \qquad (64.1)$$

According to their definition the quantities ∇v^i transform like the contravariant components of a vector, although this is evidently not true of the quantities dv^i. For this reason ∇v^i is called the *absolute differential* of v^i. $d\mathbf{v}$ is often loosely termed the absolute differential of **v**.

Partial derivatives may be introduced instead of differentials; ∇v^i can be expressed as a linear differential form with respect to dy^k as follows

$$\nabla v^i = \partial_k v^i \, dy^k + \Gamma_{k\,h}^{\;i} v^k \, dy^h = (\partial_k v^i + \Gamma_{k\,h}^{\;i} v^h) \, dy^k.$$

Now the dy^k (which take arbitrary values) are the contravariant components of a vector. Hence the quantities

$$\nabla_k v^i = \partial_k v^i + \Gamma_{k\,h}^{\;i} v^h \qquad (64.2)$$

are the components of a tensor, the index k being covariant. We call this tensor the *covariant derivative* of the vector **v**. It should be noted that, like the dv^i, the $\partial_k v^i$ do not form a tensor.

If the absolute differential or the covariant derivative of the vector **v** is identically zero, all the vectors of the field are identical; in other words the field is uniform.

(2) Suppose now that the vectors **v** of the field are defined in terms of their covariant components v_i. We propose to determine $d\mathbf{v}$ in terms of its covariant components ∇v_i. To this end we introduce an arbitrary uniform field **w** with components w^i and form the scalar product

$$\mathbf{w} \cdot \mathbf{v} = w^i v_i.$$

Since $d\mathbf{w}$ is zero, we have

$$\mathbf{w} \cdot d\mathbf{v} = w^i \, dv_i + v_i \, dw^i$$

and
$$\nabla w^i = dw^i + \omega_{\;h}^{i} w^h = 0,$$

so we obtain

$$\mathbf{w} \cdot d\mathbf{v} = w^i \, dv_i - \omega_{\;h}^{i} v_i w^h,$$

or
$$w^i \nabla v_i = w^i (dv_i - \omega_{\;i}^{h} v_h).$$

This equality holds for any values of the w^i, so we have

$$\nabla v_i = dv_i - \omega_{\;i}^{h} v_h. \qquad (64.3)$$

∇v_i is called the absolute differential of v_i. Similar reasoning to that above enables us to pass from differentials to derivatives and it follows that

$$\nabla_k v_i = \partial_k v_i - \Gamma_{k\,i}^{\;h} v_h, \qquad (64.4)$$

where the $\nabla_k v_i$ are the covariant components of the covariant tensor derivative of the vector **v**.

65. Absolute differential of a tensor. The preceding considerations may be extended without difficulty to any field of tensors of order q. Let us determine the differential of a tensor of the field, considered as a vector of the space $E_n^{(q)}$, in terms of its components referred to the natural frame (M, e_i). We shall call this the *absolute differential* of the tensor. To simplify the working we shall consider as an example a tensor **T** of order 2 defined by a system of mixed components $t_i{}^j$. The components of the absolute differential relative to (M, e_i) are written as $\nabla t_i{}^j$.

Consider two arbitrary uniform vector fields **v** and **w** and form the scalar $t_i{}^j v^i w_j$. The differential of this scalar, for an infinitesimal variation of M, is given by

$$d(t_i{}^j v^i w_j) = \nabla(t_i{}^j v^i w_j) = (t_i{}^j + \nabla t_i{}^j) v^i w_j - t_i{}^j v^i w_j$$

since the two vector fields are uniform.
On the other hand

$$d(t_i{}^j v^i w_j) = dt_i{}^j v^i w_j + t_i{}^j dv^i w_j + t_i{}^j v^i dw_j$$

where, since the fields are uniform,

$$dv^i = -\omega^i{}_h v^h, \qquad dw_j = \omega^h{}_j w_h.$$

We conclude, on relabelling the summation indices, that

$$\nabla t_i{}^j v^i w_j = (dt_i{}^j - t_h{}^j \omega^h{}_i + t_i{}^h \omega^j{}_h) v^i w_j.$$

Since this equality holds regardless of the values of v^i and w_j it follows that

$$\nabla t_i{}^j = dt_i{}^j - \omega^h{}_i t_h{}^j + \omega^j{}_h t_i{}^h. \tag{65.1}$$

The quantity $\nabla t_i{}^j$ can also be expressed as a linear differential form with respect to the dy^k. Using (60.4) we get

$$\nabla t_i{}^j = (\partial_k t_i{}^j - \Gamma_k{}^h{}_i t_h{}^j + \Gamma_h{}^j{}_k t_i{}^h) dy^k$$

which may be written as

$$\nabla t_i{}^j = \nabla_k t_i{}^j dy^k,$$

where

$$\nabla_k t_i{}^j = \partial_k t_i{}^j - \Gamma_k{}^h{}_i t_h{}^j + \Gamma_h{}^j{}_k t_i{}^h. \tag{65.2}$$

It is clear that the quantities $\nabla_k t_i{}^j$ are the components of a tensor. We call this the *covariant derivative* of **T**. The general rules of composition for the absolute differential and for the covariant derivative of a tensor may be inferred immediately from (65.1) and (65.2).

The usual rules hold for the absolute differential of a sum or product of tensors. Absolute differentiation and contraction are obviously permutable operations in the sense that the result is independent of the order in which they are carried out. It follows that contracted products differentiate according to the same rule as general tensor products. There is an important result due to Ricci concerning the absolute differential of the fundamental tensor g_{ij}. This differential is given by

$$\nabla g_{ij} = dg_{ij} - \omega^h{}_i g_{hj} - \omega^h{}_j g_{ih}.$$

Using the equations (61.2), which were used in the determination of the $\omega^j{}_i$, we see that the right-hand side is identically zero.

THEOREM: *The absolute differential of the fundamental tensor g_{ij} is zero.*

It follows immediately that absolute differentiation and the raising or lowering of indices are commuting operations.

66. Acceleration vector of a moving point. Let us consider a moving point M in the space $\mathscr{E}_n$ whose position is a function of a scalar parameter t which we shall regard initially as the time in order to be able to use the terminology of elementary kinematics. Then the curvilinear coordinates (y^i) of M are functions of t and the velocity vector of M is given by

$$\mathbf{v} = \frac{d\mathbf{M}}{dt}.$$

Since the vector $d\mathbf{M}$ has contravariant components dy^i, the vector **v** has contravariant components

$$v^i = \frac{dy^i}{dt}. \tag{66.1}$$

The acceleration vector of M is the derivative of the velocity vector

$$\gamma = \frac{d\mathbf{v}}{dt}.$$

Since $d\mathbf{v}$ has contravariant components ∇v^i, the vector γ has the contravariant components given by

$$\gamma^i = \frac{\nabla v^i}{dt} = \frac{dv^i}{dt} + \Gamma_k{}^i{}_h v^h \frac{dy^k}{dt}$$

which becomes, using (66.1),

$$\gamma^i = \frac{d^2 y^i}{dt^2} + \Gamma_k{}^i{}_h \frac{dy^h}{dt} \frac{dy^k}{dt}. \tag{66.2}$$

If the acceleration of M is always zero its trajectory is a straight line in $\mathscr{E}_n$. Hence the straight lines of $\mathscr{E}_n$ are defined in terms of curvilinear coordinates (y^i) by the system of differential equations

$$\frac{d^2 y^i}{dt^2} + \Gamma_k{}^i{}_h \frac{dy^h}{dt} \frac{dy^k}{dt} = 0 \quad (i = 1, \ldots, n). \tag{66.3}$$

A convenient choice for the independent variable t is the distance s of a point of the line measured along the line from a fixed origin. The functions $y^i(s)$ therefore satisfy the differential equations

$$\frac{d^2 y^i}{ds^2} + \Gamma_k{}^i{}_h \frac{dy^k}{ds} \frac{dy^h}{ds} = 0$$

on a straight line in $\mathscr{E}_n$. Equation (66.3) was derived by making an appeal to the existing ideas of elementary kinematics. More generally, if t represents *any* parameter it may be shown that the equation represents straight lines in $\mathscr{E}_n$. This point will be taken up again in §76.

V. DIFFERENTIAL OPERATORS IN CURVILINEAR COORDINATES

67. Gradient of a scalar function. Consider a scalar field defined by a function ϕ of the curvilinear coordinates y^i of the point M. The absolute differential of this field reduces to the ordinary differential $d\phi$ of the scalar ϕ, this differential being itself a scalar. The corresponding covariant derivatives are given by $\partial_k \phi$. It is easy to verify that the

$\partial_k \phi$ transform under a change of coordinates like the covariant components of a vector. This vector is called the *gradient* of the function ϕ. We write

$$\text{grad}_k \phi = \partial_k \phi. \tag{67.1}$$

The contravariant components of the gradient of ϕ are given by

$$\text{grad}^i \phi = g^{ik} \partial_k \phi. \tag{67.2}$$

Beltrami has introduced the *norm* of the gradient given by

$$\Delta_1 \phi = g^{ij} \partial_i \phi \, \partial_j \phi, \tag{67.3}$$

which he calls the *differential parameter of the first kind*. In elementary geometry, when the space is referred to Cartesian coordinates, this is given by

$$\Delta_1 \phi = \left(\frac{\partial \phi}{\partial x}\right)^2 + \left(\frac{\partial \phi}{\partial y}\right)^2 + \left(\frac{\partial \phi}{\partial z}\right)^2.$$

68. Curl of a vector field. Consider a vector field **v** with covariant components v_i. We have seen that

$$\nabla_j v_i = \partial_j v_i - \Gamma_j{}^k{}_i v_k. \tag{68.1}$$

Since the Γ's are symmetric in their lower indices we can interchange the indices i and j in (68.1) to get

$$\nabla_i v_j = \partial_i v_j - \Gamma_j{}^k{}_i v_k. \tag{68.2}$$

Subtracting (68.2) from (68.1) we find

$$\nabla_i v_j - \nabla_j v_i = \partial_i v_j - \partial_j v_i.$$

It follows that the quantities

$$\partial_i v_j - \partial_j v_i$$

are the covariant components of an antisymmetric tensor. This tensor is called the *curl* (or *rotor* or *rotation*) of the vector **v**. We write

$$\text{curl}_{ji} \mathbf{v} = \partial_i v_j - \partial_j v_i. \tag{68.3}$$

In elementary geometry the vector adjoint of the antisymmetric tensor (68.3) is called the curl vector.

69. Divergence of a vector field. Consider a vector field **v** with contravariant components v^i. We call the scalar

$$\operatorname{div} \mathbf{v} = \nabla_i v^i$$

the *divergence* of the vector **v**. Since

$$\nabla_j v^i = \partial_j v^i + \Gamma^i_{jh} v^h,$$

we have

$$\operatorname{div} \mathbf{v} = \partial_i v^i + \Gamma^i_{ih} v^h. \tag{69.1}$$

This formula may be rewritten using a simplified expression for the quantities Γ^i_{ih}. Ricci's theorem (§65) relating to the fundamental tensor may be written in covariant form as

$$\nabla_h g_{ij} = \partial_h g_{ij} - \Gamma^k_{hi} g_{kj} - \Gamma^k_{hj} g_{ik} = 0.$$

Contracted multiplication of this by g^{ij} gives

$$g^{ij} \partial_h g_{ij} - \Gamma^i_{hi} - \Gamma^j_{hj} = 0,$$

or

$$\Gamma^i_{ih} = \tfrac{1}{2} g^{ij} \partial_h g_{ij}. \tag{69.2}$$

The form of the right-hand side of (69.2) brings to mind the derivative of the determinant g. If α^{ij} denotes the cofactor of the element g_{ij} in g we have

$$\partial_h g = \alpha^{ij} \partial_h g_{ij} = g g^{ij} \partial_h g_{ij}.$$

It follows that

$$\Gamma^i_{ih} = \frac{1}{2} \frac{\partial_h g}{g} = \frac{\partial_h \sqrt{|g|}}{\sqrt{|g|}}. \tag{69.3}$$

Substituting (69.3) in (69.1) we obtain

$$\operatorname{div} \mathbf{v} = \frac{1}{\sqrt{|g|}} \partial_i [\sqrt{(|g|)} v^i]. \tag{69.4}$$

Elements of tensor calculus

In curvilinear coordinates the following integral over an n-dimensional domain $\mathscr{D}$ may be used to introduce the multiple integral of order n:

$$\int_{\mathscr{D}} \operatorname{div} \mathbf{v} \, dV = \int_{\mathscr{D}} \frac{1}{\sqrt{|g|}} \, \partial_i[\sqrt{(|g|)} v^i] \, \sqrt{(|g|)} \, dy^1 \ldots dy^n$$

$$= \int_{\mathscr{D}} \partial_i[\sqrt{(|g|)} v^i] \, dy^1 \ldots dy^n.$$

The last expression can be transformed into an integral over the boundary of $\mathscr{D}$ which is just the flux of the vector field $\mathbf{v}$ passing through this boundary.

70. The Laplacian of a function. In the space of elementary geometry referred to a Cartesian coordinate system (x, y, z) the expression

$$\Delta_2 \phi = \frac{\partial^2 \phi}{\partial x^2} + \frac{\partial^2 \phi}{\partial y^2} + \frac{\partial^2 \phi}{\partial z^2} = \operatorname{div} \operatorname{grad} \phi$$

is often of interest. It is called the *Laplacian* of the function ϕ (or the *Beltrami differential parameter of the second kind*).

More generally, given a scalar function ϕ of the variables y^i in $\mathscr{E}_n$, we call

$$\Delta_2 \phi = \operatorname{div} \operatorname{grad} \phi \qquad (70.1)$$

the *Laplacian* of ϕ. The preceding considerations enable us to determine an expression for the Laplacian of ϕ in terms of any system of curvilinear coordinates. Using Ricci's theorem

$$\Delta_2 \phi = \nabla_i(g^{ij} \partial_j \phi) = g^{ij} \nabla_i(\partial_j \phi).$$

It follows that

$$\Delta_2 \phi = g^{ij}(\partial_{ij} \phi - \Gamma_i{}^k{}_j \partial_k \phi). \qquad (70.2)$$

Using the expression (69.4) for the divergence we deduce the following alternative form of the Laplacian which is often convenient in practical calculations

$$\Delta_2 \phi = \frac{1}{\sqrt{|g|}} \, \partial_i[\sqrt{(|g|)} g^{ij} \partial_j \phi]. \qquad (70.3)$$

CHAPTER V

Riemannian Spaces

I. TANGENTIAL AND OSCULATING EUCLIDEAN METRICS

71. Definition of Riemannian spaces. Let us consider an n-dimensional point continuum V_n which is many times differentiable. The best way to visualize such a continuum is by imagining that it corresponds to a dynamical system with n degrees of freedom. We assume that the immediate neighbourhood of each point M_0 of V_n can be represented by a set of n coordinates y^i capable of assuming all values in the neighbourhood of the y_0^i which are the coordinates of M_0. These coordinates y^i, which serve to represent analytically a certain part of V_n, can obviously be chosen in an infinite number of ways. We shall arrange in transforming from one system of coordinates (y^i) to another $(y^{j'})$, that the new coordinates are continuously differentiable functions of the old (to a sufficient order), and conversely.

Let us associate the continuum V_n with the metric given by

$$ds^2 = g_{ij} dy^i dy^j \tag{71.1}$$

where the coefficients g_{ij} are *arbitrary* functions of the y^i, subject *only* to the condition of being continuously differentiable to a sufficiently high order. If this metric does not satisfy the integrability conditions given in §61, curvilinear coordinate systems cannot exist in $\mathscr{E}_n$ such that the metric of $\mathscr{E}_n$ takes the form (71.1). In this case we say that the metric (71.1) is non-Euclidean and that it defines V_n as a *Riemannian space*.

A Riemannian space is thus simply an n-dimensional continuum with an arbitrary metric. Such a space is said to be *properly* Riemannian if the metric is positive definite. The expression (71.1) can, quite generally, be put into the form of an algebraic sum of n squares of linear differential forms. The set of signs ($+$ or $-$) which precede the squared terms in this sum is called the *signature* of the form.

72. Tangential Euclidean metric at a point. The simplest way of describing the geometrical properties of a Riemannian space is to identify it locally, as far as possible, with a Euclidean space $\mathscr{E}_n$. To this end we introduce the concept of a tangential Euclidean metric at a point M_0 of V_n.

Consider a Euclidean space $\mathscr{E}_n$ with the same signature as V_n and a point M_0 of the Riemannian space with coordinates (y_0^i). Suppose the point M_0 to correspond to a point m_0 of $\mathscr{E}_n$ with a natural frame $(m_0, \mathbf{e}_i)$ which is subject only to the conditions

$$\mathbf{e}_i \cdot \mathbf{e}_j = (g_{ij})_0 \qquad (72.1)$$

where $(g_{ij})_0$ designates the value at M_0 of the coefficients g_{ij} of (71.1).

Suppose that each point M in the vicinity of M_0 in V_n is brought into correspondence with a point m in the vicinity of m_0 in $\mathscr{E}_n$ in the following way: if M has the coordinates (y^i) then m is defined by

$$\overrightarrow{m_0 m} = [(y^i - y_0^i) + \Psi_{(2)}^i (y^k - y_0^k)] \mathbf{e}_i, \qquad (72.2)$$

where the functions $\Psi_{(2)}^i$ are restricted to be at least of second order with respect to the variables $(y^k - y_0^k)$. We then say that the correspondence defines a *first-order representation* of the vicinity of M_0. The point m is said to be the *image* of M in the representation, m_0 being naturally the image of M_0.

According to the formula (72.2) the point m is defined as a function of the n scalar variables (y^i). It follows that the (y^i) constitute a system of curvilinear coordinates in the Euclidean space $\mathscr{E}_n$ in the neighbourhood of m_0. This system of curvilinear coordinates has the natural frame at m_0 defined by the vectors

$$\left(\frac{\partial \mathbf{m}}{\partial y^i} \right)_0 = \mathbf{e}_i$$

which coincide with the frame $(m_0, \mathbf{e}_i)$ given initially.

It follows from (72.1), if the metric of the Euclidean space $\mathscr{E}_n$ in the coordinate system (y^i) is designated by

$$\overline{ds}^2 = \bar{g}_{ij} dy^i dy^j \qquad (72.3)$$

then for $y^i = y_0^i$, we have

$$(\bar{g}_{ij})_0 = \mathbf{e}_i \cdot \mathbf{e}_j = (g_{ij})_0. \tag{72.4}$$

The metrics (71.1) and (72.3) thus have the same coefficients for $y^i = y_0^i$ and are said to be *tangential* at $y^i = y_0^i$.

Let us make a general transformation of the (y^i) to new coordinates ($y^{j'}$) and let ($m_0, \mathbf{e}_{j'}$) be the natural frame at m_0 for the coordinates ($y^{j'}$). It follows at once that the vector $\overrightarrow{m_0 m}$ defined by (72.2) can also be written as

$$\overrightarrow{m_0 m} = [(y^{j'} - y_0^{j'}) + \mathcal{E}_{(2)}^{j'}(y^{l'} - y_0^{l'})]\mathbf{e}_{j'},$$

where the functions $\mathcal{E}$ satisfy the same condition as the Ψ. It is now apparent that the *concept of first-order representation is independent of the system of variables employed* or, as we shall say, it has an *intrinsic character*.

In such a coordinate transformation the $(\bar{g}_{ij})_0$ change according to the relations

$$(\bar{g}_{ij})_0 = (A_i^{k'})_0 (A_j^{l'})_0 (\bar{g}_{k'l'})_0$$

where $$A_i^{k'} = \frac{\partial y^{k'}}{\partial y^i}.$$

In order that the concept of tangential Euclidean metric shall have an intrinsic character it is necessary and sufficient that

$$(g_{ij})_0 = (A_i^{k'})_0 (A_j^{l'})_0 (g_{k'l'})_0.$$

We are therefore led to adopt the convention that, in any coordinate transformation, the coefficients g_{ij} of the metric of a Riemannian space transform according to the relations

$$g_{ij} = A_i^{k'} A_j^{l'} g_{k'l'}. \tag{72.5}$$

Since the concepts of first-order representation and of tangential Euclidean metric at a point have an intrinsic character we can extend to Riemannian spaces some geometrical ideas which are Euclidean in origin.

73. Geometrical ideas derived from tangential Euclidean metrics.
(*1*) Consider a point M_0 in the Riemannian space V_n with the metric

$$ds^2 = g_{ij}dy^i dy^j. \tag{73.1}$$

Let $\mathscr{E}_n$ be the tangential Euclidean space at M_0 which we suppose to be referred to the natural frame (m_0, e_i) at m_0 associated with the system of curvilinear coordinates (y^i). The metric of $\mathscr{E}_n$ at m_0 is given by the quadratic form

$$(g_{ij})_0 dy^i dy^j.$$

The tensors which have components referred to the frame (m_0, e_i) in $\mathscr{E}_n$ will be said to define tensors at the point M_0 of V_n with respect to the coordinate system (y^i). On transforming from the coordinates (y^i) to the $(y^{j'})$ the frame (m_0, e_i) is replaced by $(m_0, e_{j'})$ where

$$e_i = (A_i^{j'})_0 e_{j'}, \qquad e_{j'} = (A_{j'}^i)_0 e_i, \tag{73.2}$$

and the components $t^{ij}{}_k$, for example, of a third-order tensor transform according to the usual rule

$$t^{ij}{}_k = (A_{l'}^i)_0 (A_{m'}^j)_0 (A_k^{n'})_0 t^{l'm'}{}_{n'}. \tag{73.3}$$

The components of different kinds of tensor **T** can be derived from one another by contracted multiplication with the quantities $(g_{ij})_0$ and $(g^{ij})_0$.

If, for the moment, we designate the point chosen in V_n by M we may omit the index zero from the above formulae. In this way the complete Euclidean tensor algebra can be extended without modification to vectors and tensors associated with a *single* point M of V_n. The algebra of these tensors is simply carried over from the tensor algebra of the Euclidean space $\mathscr{E}_n$ by means of the tangential Euclidean metric. In particular, denoting two vectors associated with M by **v** and **w**, and their components by v^i and w^i respectively, their scalar product is given by

$$\mathbf{v} \cdot \mathbf{w} = g_{ij} v^i w^j.$$

(2) The introduction of tangential Euclidean metrics at a point also allows us to define, in Riemannian geometry, certain ideas involving $1, \ldots, n$-dimensional domains of V_n.

Suppose, for example, that the space V_n is properly Riemannian. Using the same notation as before the infinitesimal distance between two points M_0, M in the Riemannian space is equal to the infinitesimal distance in Euclidean space between the two image points

$$\overrightarrow{m_0 m}^2 = (\bar{g}_{ij})_0 \, dy^i \, dy^j = (g_{ij})_0 \, dy^i \, dy^j = \overrightarrow{M_0 M}^2.$$

Using the infinitesimal distance between two points in the Riemannian space given by

$$ds = \sqrt{(g_{ij} \, dy^i \, dy^j)}$$

we may derive by integration the length of a finite length of arc $\overparen{AB}$ by a formula identical with (58.3), i.e.

$$\overparen{AB} = \int_a^b \sqrt{\left(g_{ij} \frac{dy^i}{dt} \frac{dy^j}{dt} \right)} \, dt.$$

Consider an element of volume at the origin M_0. This corresponds to the volume element

$$dV = \sqrt{|g_0|} \, dy^1 \dots dy^n$$

in the Euclidean space. We are thus led to adopt

$$dV = \sqrt{|g|} \, dy^1 \dots dy^n$$

as the expression for the volume element in the Riemannian space. The volume V of an n-dimensional domain will therefore be given by the integral

$$V = \int |g| \, dy^1 \dots dy^n.$$

Analogous results could be given for various kinds of area.

It should be stressed that the notion of the tangential Euclidean metric does not permit us to interrelate tensors associated with different points of Riemannian space, even neighbouring ones.

74. Osculating Euclidean metric at a point. Using the same notation as in §72, let us take a first-order representation and consider the point m, which is the image of M, determined by the curvilinear co-ordinates y^i. From (61.7) we have

$$\frac{\partial^2 \mathbf{m}}{\partial y^j \, \partial y^k} = \overline{\begin{Bmatrix} i \\ j \, k \end{Bmatrix}} \mathbf{e}_i, \tag{74.1}$$

where the Christoffel symbols on the right-hand side are constructed from the $\bar{g}_{ij}$ of the Euclidean metric. This leads us to make the following definition:

We say that a representation for the vicinity of M_0 is of *second order* if it is of first order and if the formula (72.2) is replaced by

$$\overrightarrow{m_0 m} = \left[y^i - y_0^i + \frac{1}{2} \begin{Bmatrix} i \\ j \, k \end{Bmatrix}_0 (y^j - y_0^j)(y^k - y_0^k) + \Psi^i_{(3)}(y^r - y_0^r) \right] \mathbf{e}_i, \tag{74.2}$$

where the Christoffel symbols of the right-hand side are constructed from the g_{ij} and evaluated for $y^i = y_0^i$. The functions $\Psi^i_{(3)}$ are restricted to be at least of third order in the variables $(y^r - y_0^r)$ in the vicinity of $(y^r - y_0^r) = 0$.

Consider the metric

$$ds^2 = \bar{g}_{ij} \, dy^i \, dy^j \tag{74.3}$$

of the Euclidean space $\mathscr{E}_n$ referred to the system of curvilinear coordinates (y^i) defined by means of a second-order representation. Since the representation is already of first order we have

$$\left(\frac{\partial \mathbf{m}}{\partial y^i} \right)_0 = \mathbf{e}_i,$$

and consequently

$$(\bar{g}_{ij})_0 = (g_{ij})_0.$$

It also follows, on differentiating (74.2), that

$$\left(\frac{\partial^2 \mathbf{m}}{\partial y^j \, \partial y^k} \right)_0 = \begin{Bmatrix} i \\ j \, k \end{Bmatrix}_0 \mathbf{e}_i.$$

Comparing this result with (74.1) evaluated at $y^i = y_0^i$ we obtain

$$\begin{Bmatrix} i \\ j \, k \end{Bmatrix}_0 = \overline{\begin{Bmatrix} i \\ j \, k \end{Bmatrix}}_0.$$

The Christoffel symbols of the first kind are therefore equal at $y^i = y_0^i$ and, from (62.1),

$$(\partial_k \bar{g}_{ij})_0 = (\partial_k g_{ij})_0. \tag{74.4}$$

We have just seen that the Riemannian metric and the Euclidean metric (74.3) have the same coefficients and the same derivatives of these coefficients at $y^i = y_0^i$. We can also show, as in the previous case, that the concepts of second-order representation and of the osculating metric have an intrinsic character. These concepts will be used to define the absolute differential of a vector or tensor in Riemannian space.

In order to gain a more concrete idea of this representation let us consider a simple example. Locally curved surfaces set in Euclidean space can be regarded as two-dimensional Riemannian spaces. Let S be a surface, M_0 a point on S and $\mathscr{E}_2$ the tangent plane at M_0. Let each point M of S in the vicinity of M_0 be put in correspondence with the point m which is the orthogonal projection of M on $\mathscr{E}_2$. The correspondence between M and m defines a representation of second order; the metric of $\mathscr{E}_2$ osculates with that of the surface at M_0.

75. Tensor fields in V_n. Absolute differentiation. We associate a tensor **T** with each point M of V_n in the following manner. Each point M is associated with a frame compatible with the metric at that point; the components of the tensor **T** can then be obtained with respect to such a frame. If, for each point M, this set of components is given as a function of the coordinates (y^i), we say that we have a tensor field in V_n. The g_{ij} constitute such a tensor field.

In order to compare tensors associated with two neighbouring points M_0 and M of V_n it is necessary to inter-relate the corresponding frames in $\mathscr{E}_n$. To this end we use a second-order representation in $\mathscr{E}_n$ in the neighbourhood of M_0 and adopt the natural frames associated with the points m_0 and m which are images of M_0 and M. The tensor associated with the point M is represented by the image tensor defined by the components referred to the natural frame at m.

When we change the second-order representation, the functions $\Psi^i_{(3)}$ are modified and the point m is replaced by a point which differs from it only by a third-order infinitesimal vector, the $(y^i - y_0^i)$ being the first-order infinitesimals. The vectors of the natural frame at m are modified by vectors which are second-order infinitesimals or less. It follows that the image tensor of the tensor associated with the point M is defined to the same order of approximation.

Let $\mathbf{T}_0$ and $\mathbf{T}$ be two tensors associated with the points M_0 and M respectively. They have, in any second-order representation, two image tensors whose difference is defined independently of the particular representation considered to the second order in the infinitesimals or less. The principal part of this difference between the image tensors is called the *absolute differential* of the tensor $\mathbf{T}$.

Consider, for example, a field of vectors in V_n defined by their contravariant components v^i. In a second-order representation the vector field has image vectors, corresponding to the points M_0 and M, whose difference has contravariant components with respect to the natural frame at m_0 given by

$$(\nabla v^i)_0 = (dv^i)_0 + (\omega^i{}_h)_0 v_0^h,$$

where

$$(\omega^i{}_h)_0 = \overline{\begin{Bmatrix} i \\ k\ h \end{Bmatrix}} dy^k. \tag{75.1}$$

The Christoffel symbols on the right-hand side of this relation are those of the osculating Euclidean metric and are therefore equal to the Christoffel symbols of the Riemannian metric at $y^i = y_0^i$.

Suppressing the index 0, we see that since

$$\omega^i{}_h = \begin{Bmatrix} i \\ k\ h \end{Bmatrix} dy^k = \Gamma_k{}^i{}_h dy^k, \tag{75.2}$$

then the absolute differential of the vector $\mathbf{v}$ at M has the contravariant components

$$\nabla v^i = dv^i + \omega^i{}_h v^h. \tag{75.3}$$

As in Euclidean geometry the quantities

$$\nabla_k v^i = \partial_k v^i + \Gamma_k{}^i{}_h v^h \tag{75.4}$$

(where the $\Gamma_k{}^i{}_h$ are as defined above) are the components of a tensor called the covariant derivative of the vector $\mathbf{v}$. If the vector field is defined by the covariant components v_i its absolute differential has the covariant components

$$\nabla v_i = dv_i - \omega^h{}_i v_h$$

and the corresponding components of the covariant derivative are given by

$$\nabla_k v_i = \partial_k v_i - \Gamma_k{}^h{}_i v_h.$$

The vectors at two neighbouring points M and M' are said to be identical if their image vectors in a second-order representation are identical; the absolute differential ∇v^i corresponding to the passage from the first vector to the second is then zero.

The formulae which define the absolute differential or the covariant derivative of any tensor may also be extended to Riemannian geometry using arguments similar to those above.

To summarize, the concepts of second-order representation and of the osculating Euclidean metric permit the extension to Riemannian spaces of the Euclidean tensor analysis relating to tensors associated with neighbouring points. This holds for all the differential operators we have studied. It is important to note, however, in the case of vectors (for example), that when the absolute differential in Euclidean geometry is an exact differential satisfying the usual integrability conditions, there is no reason to suppose that this will also be true in Riemannian geometry.

76. Acceleration vector of a moving point in V_n. Geodesics. In §66 we considered the motion of a point in Euclidean space. Let us now consider a moving point M in V_n whose position is a function of a parameter t which we shall interpret as the time. The velocity vector of M has the contravariant components

$$v^i = \frac{dy^i}{dt}$$

and its acceleration vector γ has the components

$$\gamma^i = \frac{\nabla v^i}{dt} = \frac{d^2 y^i}{dt^2} + \Gamma_k{}^i{}_h \frac{dy^h}{dt} \frac{dy^k}{dt}. \tag{76.1}$$

If the acceleration of M is zero its trajectory is called a *geodesic* of the Riemannian space V_n. Generally geodesics are defined parametrically, for *any* t, by the solutions of the system of differential equations

$$\frac{d^2 y^i}{dt^2} + \Gamma_k{}^i{}_h \frac{dy^h}{dt} \frac{dy^k}{dt} = 0 \quad (i = 1, 2, \dots, n). \tag{76.2}$$

Take as independent variable along a geodesic C the distance s measured from some fixed origin of a point M along C. The vector **u** with components

$$u^i = \frac{dy^i}{ds}$$

is thus the unit vector collinear with $d\mathbf{M}$ and tangential to the curve. The geodesics are characterized by

$$\frac{\nabla u^i}{ds} \equiv \frac{dy^k}{ds} \nabla_k u^i \equiv u^k \nabla_k u^i = 0. \tag{76.3}$$

These equations show that transport along C between neighbouring points leaves the vector **u** *unchanged. Geodesics are therefore the analogues in Riemannian geometry of straight lines in Euclidean space.*

However, in $\mathscr{E}_n$ straight lines are characterized by the following property: a straight line between two points A and B corresponds to an extremum of the length of arc $\widehat{AB}$ with respect to the lengths of arc of all curves joining A and B. We may ask whether this is also true for the geodesics of Riemannian geometry. For simplicity we shall study a proper Riemannian space.

Consider, in terms of some parametric representation, an arc joining two points A and B of V_n. If a and b represent the values of the parameter t at A and B, then from §73 the length of arc $\widehat{AB}$ is given by the integral

$$\int_a^b \sqrt{[f(y^k, y'^i)]}\, dt,$$

where

$$f = g_{ij} y'^i y'^j, \qquad y'^i = \frac{dy^i}{dt}.$$

The curves representing the extrema of this integral, *the extremals*, are defined by the appropriate system of Euler's equations in the calculus of variations,

$$\frac{d}{dt}\left(\frac{\partial \sqrt{f}}{\partial y'^i}\right) - \frac{\partial \sqrt{f}}{\partial y^i} = 0 \quad (i = 1, 2, \ldots, n).$$

Carrying out the differentiations we obtain

$$\frac{d}{dt}\left[\frac{1}{\sqrt{f}}\frac{\partial f}{\partial y'^i}\right] - \frac{1}{\sqrt{f}}\frac{\partial f}{\partial y^i} = 0$$

or

$$\frac{d}{dt}\left(\frac{\partial f}{\partial y'^i}\right) - \frac{\partial f}{\partial y^i} - \frac{1}{2f}\frac{df}{dt}\frac{\partial f}{\partial y'^i} = 0. \tag{76.4}$$

Since the parameter t is still arbitrary we take this to be the distance s along the curves considered. Under these conditions

$$y'^i = \frac{dy^i}{ds}, \qquad f = g_{ij}y'^i y'^j = 1 \tag{76.5}$$

and hence the system of differential equations (76.4) reduces to

$$\frac{d}{ds}\left(\frac{\partial f}{\partial y'^i}\right) - \frac{\partial f}{\partial y^i} = 0, \tag{76.6}$$

which are Euler's equations for the function f. We deduce from (76.5) that

$$\frac{\partial f}{\partial y'^i} = 2g_{ij}y'^j, \qquad \frac{\partial f}{\partial y^i} = \partial_i g_{jk}y'^j y'^k.$$

Hence (76.6) takes the explicit form

$$\frac{d}{ds}(g_{ij}y'^j) - \tfrac{1}{2}\partial_i g_{jk}y'^j y'^k = g_{ij}\frac{dy'^j}{ds} + (\partial_k g_{ij} - \tfrac{1}{2}\partial_i g_{jk})y'^j y'^k = 0 \tag{76.7}$$

and, on introducing the Christoffel symbols of the first kind

$$g_{ij}\frac{dy'^j}{ds} + [jk, i]y'^j y'^k = 0. \tag{76.8}$$

After contracted multiplication by g^{ih} this becomes

$$\frac{d^2 y^h}{ds^2} + \Gamma_{jk}^{h} \frac{dy^j}{ds} \frac{dy^k}{ds} = 0.$$

which coincides with the geodesic equation (76.2) when $t = s$. We can therefore formulate the following THEOREM: *In Riemannian geometry, geodesics are the curves defined by the extremals of the integral representing the length of arc of any curve connecting two fixed points of V_n.*

II. THE TRANSPORT OF EUCLIDEAN METRICS

77. Mapping a curve of V_n on Euclidean space. Consider any curve C of V_n which is described parametrically. The coordinates y^i of a point M of C are functions of some parameter t. Denote a certain point on C by A; for example, that point which corresponds to the value $t = 0$.

Each point M of C is made to correspond to a point m with the natural frame (m, e_i) in the Euclidean space $\mathscr{E}_n$ in the following way:

(*1*) To the point A there corresponds an arbitrarily chosen point a with the frame $[a, (e_i)_0]$ whose orientation is not determined. The magnitudes and relative inclinations of the $(e_i)_0$ are, however, well defined by the relations

$$(e_i)_0 \cdot (e_j)_0 = (g_{ij})_0 \tag{77.1}$$

where the $(g_{ij})_0$ are the coefficients of the Riemannian metric at the point A $(t = 0)$.

(2) The point m and the vectors e_i satisfy the equations

$$\frac{d\mathbf{m}}{dt} = \frac{dy^i}{dt} \mathbf{e}_i, \qquad \frac{d\mathbf{e}_i}{dt} = (\Gamma_{ki}^{h})_M \frac{dy^k}{dt} \mathbf{e}_h, \tag{77.2}$$

where the $(\Gamma_{ki}^{h})_M$ are the values of the Christoffel symbols at the point M and are consequently functions of the single parameter t.

The integration of the system of differential equations (77.2) with the initial conditions (*1*) defines a point m and a frame (m, e_i) for each value of t and similarly for every point M of C. The path γ in $\mathscr{E}_n$

followed by m when M describes C is called the *mapping* (or *development*) of C on the Euclidean space. If the initial conditions (*1*) are varied the initial frame $[a, (e_i)_0]$ is changed and so is γ.

As an example of such a mapping imagine a curve C to be traced on a surface S and apply a developable surface to S along C. If this surface is now laid on a plane it defines a plane curve γ which can be thought of as a flat map of C.

78. The transport of Euclidean metrics along a curve. We wish to establish the following fundamental theorem concerning the mapping γ of C:

It is possible to find a metric in $\mathscr{E}_n$ such that the numerical values of its coefficients and their first derivatives along γ coincide with the numerical values of the coefficients of the Riemannian metric and their first derivatives at the corresponding points of C. In other words, it is possible to construct a Euclidean metric which osculates with the Riemannian metric simultaneously for all points of C.

In order to simplify the notation we make a coordinate transformation such that C is defined by the equations

$$y^1 = y^2 = \ldots = y^{n-1} = 0,$$

and take y^n to be the parameter t. It will be convenient to use Greek indices to represent the range of values $(1, 2, \ldots, n-1)$ and Latin indices to represent $(1, 2, \ldots, n-1, n)$. With these conventions the equations (77.2) which determine the mapping γ of C take the form

$$\frac{d\mathbf{m}}{dy^n} = \mathbf{e}_n, \qquad \frac{d\mathbf{e}_i}{dy^n} = (\Gamma_{ni}^{h})_{y^\alpha=0}\mathbf{e}_h. \tag{78.1}$$

Suppose that a correspondence is set up between each point P of V_n in the vicinity of M and a point p of $\mathscr{E}_n$ in the vicinity of m in the following way: if P has the coordinates y^i and m is the point of γ whose parameter is y^n, then the point p is defined by

$$\overrightarrow{mp} = y^\lambda \mathbf{e}_\lambda + [\tfrac{1}{2}(\Gamma_{\lambda\mu}^{i})_{y^\alpha=0}y^\lambda y^\mu + \Psi^i_{(3)}(y^\lambda)]\mathbf{e}_i, \tag{78.2}$$

where the $\Psi^i_{(3)}$ are of third order in the y^λ.

Equation (78.2) defines the point p of $\mathscr{E}_n$ as a function of the n scalar variables (y^i). The (y^i) therefore constitute a system of curvilinear coordinates for $\mathscr{E}_n$ in the vicinity of γ. The natural frame at $m(y^\alpha = 0)$ for this system of curvilinear coordinates is, from (78.1) and (78.2), defined by the vectors

$$\left(\frac{\partial \mathbf{p}}{\partial y^\lambda}\right)_{y^\alpha = 0} = \mathbf{e}_\lambda, \qquad \left(\frac{\partial \mathbf{p}}{\partial y^n}\right)_{y^\alpha = 0} = \frac{d\mathbf{m}}{dy^n} = \mathbf{e}_n,$$

which is to say that it coincides with the frame $(m, \mathbf{e}_i)$ obtained in the mapping. The metric of $\mathscr{E}_n$ in the coordinate system (y^i) has the scalar products $\mathbf{e}_i \cdot \mathbf{e}_j$ as coefficients at m. Using (77.2) or (78.1) the quantities $\mathbf{e}_i \cdot \mathbf{e}_j$ satisfy the equations

$$d(\mathbf{e}_i \cdot \mathbf{e}_j) = [(\Gamma_k{}^h{}_i)_M(\mathbf{e}_j \cdot \mathbf{e}_h) + (\Gamma_k{}^h{}_j)_M(\mathbf{e}_i \cdot \mathbf{e}_h)]\, dy^k \qquad (78.3)$$

as M describes C.

On the other hand, according to the definition of the Christoffel symbols, the Riemannian metric coefficients g_{ij} satisfy

$$dg_{ij} = (\Gamma_{kij} + \Gamma_{kji})\, dy^k,$$

therefore

$$dg_{ij} = [(\Gamma_k{}^h{}_i)_M g_{jh} + (\Gamma_k{}^h{}_j)_M g_{ih}]\, dy^k. \qquad (78.4)$$

The quantities $\mathbf{e}_i \cdot \mathbf{e}_j$ and g_{ij} thus satisfy the same differential system as the point M describes C. Moreover, from (77.1), the initial conditions at A are the same. It follows that

$$\mathbf{e}_i \cdot \mathbf{e}_j = g_{ij}$$

identically as M describes C. The Euclidean metric and the Riemannian metric are therefore tangential at all points of C.

To demonstrate that they are osculating it is now sufficient to establish that the quantities $(\Gamma_i{}^h{}_j)_{y^\alpha = 0}$ are the values of the Christoffel symbols on γ for the Euclidean metric. These symbols are the coefficients of $\mathbf{e}_k$ in $(\partial^2 \mathbf{p})/(\partial y^i \partial y^j)$. Now, using (78.2),

$$\left(\frac{\partial^2 \mathbf{p}}{\partial y^\lambda \partial y^\mu}\right)_{y^\alpha = 0} = (\Gamma_\lambda{}^h{}_\mu)_{y^\alpha = 0}\mathbf{e}_h,$$

and from (78.1),

$$\left(\frac{\partial^2 \mathbf{p}}{\partial y^i \, \partial y^n}\right)_{y^\alpha = 0} = \frac{d}{dy^n}\left(\frac{\partial \mathbf{p}}{\partial y^i}\right)_{y^\alpha = 0} = \frac{d\mathbf{e}_i}{dy^n} = (\Gamma_n{}^h{}_i)_{y^\alpha = 0} \, \mathbf{e}_h,$$

which completes the demonstration of our theorem. The Euclidean metric obtained will be called the *Euclidean metric transported along* C.

79. Geometrical applications. A number of geometrical results can be obtained from the concepts of mapping and the transport of Euclidean metrics. We limit ourselves to a few of these.

Let C be a geodesic in the Riemannian space V_n. Each point M of C is associated with the unit vector $\mathbf{u}$ tangential to C and, when we pass from the point M to a neighbouring point of C the absolute differential of $\mathbf{u}$ is zero. Now this absolute differential is the same as that of the vector image in $\mathscr{E}_n$, assuming a representation of second order in the vicinity of M (for example, that which defines the transport of a Euclidean metric). The image vector is therefore simply the unit tangent vector to the curve γ which is the mapping of C. It follows that the curve γ is a straight line in $\mathscr{E}_n$. We may therefore state the THEOREM:

The geodesics of a Riemannian space are those curves which map on Euclidean space as straight lines.

Using this theorem it is easy to deduce, in a purely geometric fashion, that the geodesics of Riemannian space are the curves which define the extreme length of arc between two fixed points, but we only mention this in passing.

Consider now any curve C of V_n and suppose that a vector $\mathbf{v}(M)$ is associated with each point M of C and is continuously variable as M describes C. The absolute differential of $\mathbf{v}$ on passing from the point M to a neighbouring point is equal to the difference between the image vectors in $\mathscr{E}_n$ for the representation associated with a Euclidean metric transported along C. We are therefore led to measure the *finite geometrical variation* of the vector $\mathbf{v}(M)$, on passing from a point A to a point B along the curve, as the difference between the image vectors of $\mathbf{v}(A)$ and $\mathbf{v}(B)$.

Suppose a vector field $\mathbf{v}(M)$ to be given in V_n. The finite geometric variation of a vector of the field in passing between two points A and B is, according to the above definition, dependent on the path C in V_n taken between A and B.

III. CURVATURE TENSOR OF A RIEMANNIAN SPACE

80. Mapping a quasi-parallelogram. In our study of Euclidean space using curvilinear coordinates it was pointed out that, in order that a differential quadratic form might define a Euclidean metric, it is necessary for the conditions of integrability of the de_i to be satisfied. In Riemannian geometry these conditions do not, in general, hold. We can, however, interpret the de_i geometrically using a method due to Cartan.

In order to evaluate the second derivatives of a scalar or vector function of the variables (y^i) it is necessary to carry out two successive differentiations which correspond to distinct variations of the y^i. We generalize this procedure by introducing two distinct differentiation symbols.

Starting with any values y^i of the variables we introduce the arbitrary variations dy^i. Let these variations define a differentiation symbol d. In a similar fashion let δy^i be a second set of arbitrary variations defining another differentiation symbol δ. Thus, starting from the values $(y^i + dy^i)$ of the variables and effecting the δ differentiation, we obtain the values

$$(y^i + dy^i + \delta y^i + \delta dy^i). \tag{80.1}$$

If, starting with the values $(y^i + \delta y^i)$, we perform the differentiation d, we likewise obtain the values

$$(y^i + \delta y^i + dy^i + d\delta y^i). \tag{80.2}$$

These two sets of values are identical if we arrange that

$$d\delta y^i = \delta dy^i. \tag{80.3}$$

We then say that the two differentiation symbols commute for scalar functions or, more briefly, *commute*. If $f(y^i)$ denotes a twice continuously differentiable function of the y^i then

$$\delta f = \partial_i f \, \delta y^i$$

and
$$d\delta f = \partial_{ij} f \, \delta y^i dy^j + \partial_i f \, d\delta y^i.$$

Using the symmetry of the second derivatives and (80.3) we have

$$d\delta f = \delta df. \qquad (80.4)$$

We note, in particular, that the commutation property of the two differentiation symbols holds for an arbitrary change of variables y^i.

Consider two interchangeable differentiation symbols and, starting from the point M of V_n with coordinates y^i, carry out the differentiation d which changes $M(y^i)$ to $M + d\mathbf{M}(y^i + dy^i)$ and the differentiation δ which changes M to $M + \delta\mathbf{M}(y^i + \delta y^i)$. We suppose that $d\mathbf{M}$ and $\delta\mathbf{M}$ are not collinear, which is equivalent to saying that dy^i and δy^i are not proportional. Starting from $M + d\mathbf{M}$ perform the differentiation δ which carries it to the point M' with coordinates (80.1). Because of the interchangeability of d and δ, we obtain the same point M' as if we had performed the differentiation d starting from $M + \delta\mathbf{M}$, since the coordinates (80.2) of the point reached coincide with (80.1). The closed path defined by the four points $(M, M + d\mathbf{M}, M', M + \delta\mathbf{M})$ will be called a *quasi-parallelogram*.

We now propose to map the alternative paths in the quasi-parallelogram between M and M' on the Euclidean space $\mathcal{E}_n$. The point M corresponds to a frame $(m, \mathbf{e}_i)$ in $\mathcal{E}_n$. If we carry out first the differentiation d and then the differentiation δ, we map the sides $(M, M + d\mathbf{M})$ and $(M + d\mathbf{M}, M')$ of the quasi-parallelogram. The frames $(m, \mathbf{e}_i)$ and $(m + d\mathbf{m}, \mathbf{e}_i + d\mathbf{e}_i)$ are related by (77.2). Hence

$$d\mathbf{m} = dy^i \mathbf{e}_i, \qquad d\mathbf{e}_i = \omega^h{}_i(d) \, \mathbf{e}_h, \qquad (80.5)$$

where $\omega^h{}_i(d)$ denotes the differential form $\omega^h{}_i$ evaluated at $(y^i + dy^i)$. Mapping the side $(M + d\mathbf{M}, M')$ then carries us from the frame $(m + d\mathbf{m}, \mathbf{e}_i + d\mathbf{e}_i)$ to the frame $[m'_1, (\mathbf{e}'_i)_1]$ with

$$\left.\begin{aligned}
\overrightarrow{mm'_1} &= d\mathbf{m} + \delta\mathbf{m} + \delta d\mathbf{m}, \\
(\mathbf{e}'_i)_1 - \mathbf{e}_i &= d\mathbf{e}_i + \delta\mathbf{e}_i + \delta d\mathbf{e}_i.
\end{aligned}\right\}$$

Reversing the order of differentiation we pass first from the frame $(m, \mathbf{e}_i)$ to the frame $(m + \delta\mathbf{m}, \mathbf{e} + \delta\mathbf{e}_i)$ with

$$\delta\mathbf{m} = \delta y^i \mathbf{e}_i, \qquad \delta\mathbf{e}_i = \omega^h{}_i(\delta) \mathbf{e}_h; \qquad (80.5')$$

and then from this frame to $[m_2', (\mathbf{e}_i')_2]$ with

$$\begin{aligned}
\overrightarrow{mm_2'} &= \delta\mathbf{m} + d\mathbf{m} + d\delta\mathbf{m}, \\
(\mathbf{e}_i')_2 - \mathbf{e}_i &= \delta\mathbf{e}_i + d\mathbf{e}_i + d\delta\mathbf{e}_i.
\end{aligned}\Bigg\}$$

It follows that $[m_1', (\mathbf{e}_i')_1]$ to $[m_2', (\mathbf{e}_i')_2]$ are related by the formulae

$$\overrightarrow{m_1' m_2'} = d\delta\mathbf{m} - \delta d\mathbf{m}, \qquad (80.6)$$

$$(\mathbf{e}_i')_2 - (\mathbf{e}_i')_1 = d\delta\mathbf{e}_i - \delta d\mathbf{e}_i, \qquad (80.7)$$

where we wish to evaluate the right-hand sides. We have

$$d\delta\mathbf{m} - \delta d\mathbf{m} = d(\delta y^i \mathbf{e}_i) - \delta(dy^i \mathbf{e}_i) = \delta y^i d\mathbf{e}_i - dy^i \delta\mathbf{e}_i,$$

and from (80.5) and (80.5')

$$d\delta\mathbf{m} - \delta d\mathbf{m} = (\Gamma_k{}^h{}_i dy^k \delta y^i - \Gamma_k{}^h{}_i dy^i \delta y^k) \mathbf{e}_h.$$

Interchanging the indices i and k in the second term on the right-hand side and using the symmetry of the Christoffel symbols in their lower indices we find that

$$d\delta\mathbf{m} - \delta d\mathbf{m} = (\Gamma_k{}^h{}_i - \Gamma_i{}^h{}_k) dy^k \delta y^i \mathbf{e}_h = 0. \qquad (80.8)$$

The two mappings thus lead to the same origin m' for the final frame. This is essentially due to the fact that the Christoffel symbols are determined in such a way that the integrability conditions are satisfied.

Now compare the vectors of the two frames with origin at m'. We have

$$\begin{aligned}
d\delta\mathbf{e}_i - \delta d\mathbf{e}_i &= d[\omega^h{}_i(\delta) \mathbf{e}_h] - \delta[\omega^h{}_i(d) \mathbf{e}_h] \\
&= [d\omega^h{}_i(\delta) - \delta\omega^h{}_i(d)] \mathbf{e}_h + \omega^k{}_i(\delta) d\mathbf{e}_k - \omega^k{}_i(d) \delta\mathbf{e}_k,
\end{aligned}$$

and using (80.5) and (80.5')

$$d\delta\mathbf{e}_i - \delta d\mathbf{e}_i = [d\omega^h{}_i(\delta) - \delta\omega^h{}_i(d) + \omega^k{}_i(\delta) \omega^h{}_k(d) - \omega^k{}_i(d) \omega^h{}_k(\delta)] \mathbf{e}_h.$$

We write
$$d\delta e_i - \delta d e_i = \Omega^h{}_i e_h \tag{80.9}$$
with
$$\Omega^h{}_i = d\omega^h{}_i(\delta) - \delta\omega^h{}_i(d) + \omega^k{}_i(\delta)\,\omega^h{}_k(d) - \omega^k{}_i(d)\,\omega^h{}_k(\delta). \tag{80.10}$$

For an arbitrary Riemannian metric, the quantities $\Omega^h{}_i$ are, in general, different from zero and the frames associated with the two mappings have different orientations; however, they have the same form and magnitude since the scalar products of the vectors of these frames are given by the coefficients of the metric at M'. The $\Omega^h{}_i$ thus define the rotation at m' that is required to pass from one frame to the other.

It is important to realize that the $\Omega^h{}_i$ are the components of a tensor. A change of curvilinear coordinates corresponds to a change of frame defined by
$$e_i = A_i^{j'} e_{j'},$$
from which we deduce that
$$\delta e_i = A_i^{j'}\,\delta e_{j'} + \delta A_i^{j'}\,e_{j'}$$
and
$$d\delta e_i = A_i^{j'}\,d\delta e_{j'} + dA_i^{j'}\,\delta e_{j'} + \delta A_i^{j'}\,d e_{j'} + d\delta A_i^{j'}\,e_{j'}.$$

On interchanging the two differentiation symbols and taking the difference between corresponding terms, it follows, since d and δ are interchangeable when acting upon the $A_i^{j'}$,
$$d\delta e_i - \delta d e_i = A_i^{j'}(d\delta e_{j'} - \delta d e_{j'}),$$
and on introducing the $\Omega^h{}_i$ as in (80.9)
$$\Omega^h{}_i e_h = A_i^{j'}\Omega^{l'}{}_{j'}\,e_{l'} = A_i^{j'}A_{l'}^{h}\Omega^{l'}{}_{j'}\,e_h.$$

Identifying the coefficients of e_h on each side we obtain the tensor transformation
$$\Omega^h{}_i = A_i^{j'}A_{l'}^{h}\Omega^{l'}{}_{j'}.$$

We therefore state the following THEOREM:

The quantities $\Omega^h{}_i$ are the components of a mixed tensor.

81. The Riemann-Christoffel tensor. It can be shown that the quantities $\Omega^h{}_i$ are bilinear forms with respect to the dy^r and δy^s. The coefficients of these forms can be evaluated in terms of the Christoffel symbols. We have

$$d\omega^h{}_i(\delta) = d(\Gamma_s{}^h{}_i \delta y^s) = \partial_r \Gamma_s{}^h{}_i dy^r \delta y^s + \Gamma_s{}^h{}_i d\delta y^s,$$

hence

$$d\omega^h{}_i(\delta) - \delta\omega^h{}_i(d) = \partial_r \Gamma_s{}^h{}_i dy^r \delta y^s - \partial_r \Gamma_s{}^h{}_i \delta y^r dy^s,$$

and on interchanging the indices r and s in the second term on the right-hand side

$$d\omega^h{}_i(\delta) - \delta\omega^h{}_i(d) = (\partial_r \Gamma_s{}^h{}_i - \partial_s \Gamma_r{}^h{}_i) dy^r \delta y^s.$$

Furthermore

$$\omega^l{}_i(\delta) \, \omega^h{}_l(d) - \omega^l{}_i(d) \, \omega^h{}_l(\delta) = (\Gamma_s{}^l{}_i \Gamma_r{}^h{}_l - \Gamma_r{}^l{}_i \Gamma_s{}^h{}_l) dy^r \delta y^s.$$

We conclude from (80.10) that the $\Omega^h{}_i$ can be expressed as the bilinear form

$$\Omega^h{}_i = R_i{}^h{}_{,rs} dy^r \delta y^s, \qquad (81.1)$$

where

$$R_i{}^h{}_{,rs} = \partial_r \Gamma_s{}^h{}_i - \partial_s \Gamma_r{}^h{}_i + \Gamma_r{}^h{}_l \Gamma_s{}^l{}_i - \Gamma_s{}^h{}_l \Gamma_r{}^l{}_i. \qquad (81.2)$$

Since dy^r and δy^s are the contravariant components of two arbitrary vectors and the $\Omega^h{}_i$ are components of a tensor, it follows that the quantities (81.2) define a tensor field in V_n. The tensor $R_i{}^h{}_{,rs}$, which is obviously antisymmetric with respect to the indices r and s, is called the *Riemann-Christoffel tensor* or *curvature tensor* of the Riemannian space V_n. The curvature of a Riemannian space V_n is thus characterized by the fact that if two distinct paths with the same extremities are mapped on Euclidean space starting with the same initial frame, then the final frames have a different orientation.

It follows from (80.9) and (81.1) that the conditions for the integrability of vectors may be expressed by the equations

$$R_i{}^h{}_{,rs} = 0. \qquad (81.3)$$

Given an arbitrary quadratic differential form, the conditions (81.3) have to be satisfied for it to be a metric in Euclidean space. It can be shown that, when the corresponding manifold is topologically equivalent to Euclidean space, these conditions are also sufficient to make V_n Euclidean; when this is not so, the Riemannian space for which the conditions (81.3) are satisfied is said to be *locally Euclidean*.† Such a space does not differ from a Euclidean space in its local properties.

82. Covariant components of the Riemann-Christoffel tensor. In order to carry out the calculation of the covariant components of the curvature tensor in a simple manner we adopt the following purely symbolic convention: an index which the covariant differential operator ∇_r does not act upon is said to be *mute*, and is written in parenthesis. With this notation we can write

$$\nabla_r \Gamma_{(i}{}^h{}_{s)} = \partial_r \Gamma_i{}^h{}_s + \Gamma_r{}^h{}_l \Gamma_i{}^l{}_s$$

and consequently

$$R_i{}^h{}_{,rs} = \nabla_r \Gamma_{(i}{}^h{}_{s)} - \nabla_s \Gamma_{(i}{}^h{}_{s)}.$$

It follows, using Ricci's theorem, that

$$R_{ij,rs} = g_{jh} R_i{}^h{}_{,rs} = \nabla_r [g_{jh} \Gamma_{(i}{}^h{}_{s)}] - \nabla_s [g_{jh} \Gamma_{(i}{}^h{}_{r)}],$$

or $\qquad R_{ij,rs} = \nabla_r \Gamma_{(i)\,j(s)} - \nabla_s \Gamma_{(i)\,j(r)}.$

Substituting the explicit form of the terms on the right-hand side

$$R_{ij,rs} = \partial_r \Gamma_{ijs} - \partial_s \Gamma_{ijr} - \Gamma_r{}^l{}_j \Gamma_{ils} + \Gamma_s{}^l{}_j \Gamma_{ilr}. \tag{82.1}$$

The second derivatives of the g_{ij} only occur in the first two terms of the right-hand side of (82.1). Let us write these terms explicitly:

$$\partial_r \Gamma_{ijs} = \partial_r [is,j] = \tfrac{1}{2} \partial_r (\partial_s g_{ij} + \partial_i g_{js} - \partial_j g_{is}),$$

$$\partial_s \Gamma_{ijr} = \partial_s [ir,j] = \tfrac{1}{2} \partial_s (\partial_r g_{ij} + \partial_i g_{jr} - \partial_j g_{ir}),$$

† For example, the surface of a cylinder set in $\mathscr{E}_3$ is locally Euclidean, but it is not topologically equivalent (i.e. homeomorphic) to a plane. (T.)

which give by subtraction,

$$R_{ij,\,rs} = \tfrac{1}{2}(\partial_{ir}g_{js}+\partial_{js}g_{ir}-\partial_{jr}g_{is}-\partial_{is}g_{jr})-g^{lm}(\Gamma_{rmj}\Gamma_{ils}-\Gamma_{smj}\Gamma_{ilr}).$$
(82.2)

Certain symmetry relations can be seen from equations (82.2). These appear even more clearly in a conveniently chosen coordinate system.

83. Normal coordinates. Relations between the components of the curvature tensor. Given a point M of V_n, carry out a change of variables from (y^i) to (z^i) which are such that the Christoffel symbols vanish and the g_{ij} remain unaltered. Then, in a second-order representation, the Euclidean space $\mathscr{E}_n$ will be referred to a system of *rectilinear* coordinates associated with the natural frame $(m, \mathbf{e}_i)$ at M. Since the natural frames at M for the (y^i) and the (z^i) are the same, the components of a given tensor at M are the same in both systems of coordinates. The coordinate system (z^i) is called the *normal coordinate system* at M associated with the coordinates (y^i). The use of such normal coordinate systems is often valuable in avoiding lengthy calculations in Riemannian geometry. We shall use them to establish certain relations between the components of the Riemann-Christoffel tensor.

The covariant components at M of the curvature tensor (which have the same values in the two coordinate systems) are given in terms of the normal coordinates at M by the relations (82.2) with the Christoffel symbols omitted:

$$R_{ij,\,rs} = \tfrac{1}{2}(\partial_{ir}g_{js}+\partial_{js}g_{ir}-\partial_{jr}g_{is}-\partial_{is}g_{jr}),$$
(83.1)

the quantities on the right-hand side being evaluated with respect to the normal coordinates. We deduce that, for any coordinate system,†

$$R_{ij,\,rs} = R_{rs,\,ij},$$
(83.2)

$$R_{ij,\,rs} = -R_{ij,\,sr} = -R_{ji,\,rs}.$$
(83.3)

† These equations certainly hold in a normal coordinate system and, being tensor equations, they must also hold in any other coordinate system. (T.)

Also, permuting the indices j, r, s, we obtain

$$R_{ir,\,sj} = \tfrac{1}{2}(\partial_{is}g_{jr} + \partial_{jr}g_{is} - \partial_{rs}g_{ij} - \partial_{ij}g_{rs}),$$
$$R_{is,\,jr} = \tfrac{1}{2}(\partial_{ij}g_{rs} + \partial_{rs}g_{ij} - \partial_{js}g_{ir} - \partial_{ir}g_{js}),$$

and adding these results

$$R_{ij,\,rs} + R_{ir,\,sj} + R_{is,\,jr} = 0. \tag{83.4}$$

Equations (83.3) and (83.4) form a complete set of identities for the components of the Riemann-Christoffel tensor: each identity connecting the components $R_{ij,\,rs}$ is satisfied by a set of numbers $R_{ij,\,rs}$ restricted only by the conditions (83.3), (83.4). Every other identity is an algebraic consequence of (83.3) and (83.4) – this is true in particular of (83.2).

84. Second-order covariant derivatives of a vector. Given a vector field in terms of the contravariant components v^h we wish to evaluate the difference between $\nabla_r(\nabla_s v^h)$ and $\nabla_s(\nabla_r v^h)$. In Euclidean space this difference is zero but in Riemannian geometry it depends upon the curvature of the space. Consider a system of normal coordinates at an arbitrary point M of V_n. By definition, at any point of V_n

$$\nabla_s v^h = \partial_s v^h + \Gamma_{s\,i}^{\,h} v^i.$$

Since the coordinates are normal at M, the Christoffel symbols vanish there and we have

$$\nabla_r(\nabla_s v^h) = \partial_{rs} v^h + \partial_r \Gamma_{s\,i}^{\,h} v^i.$$

It follows that

$$\nabla_r(\nabla_s v^h) - \nabla_s(\nabla_r v^h) = (\partial_r \Gamma_{s\,i}^{\,h} - \partial_s \Gamma_{r\,i}^{\,h}) v^i.$$

However, in a normal coordinate system we have, from (81.2),

$$R_{i\,,\,rs}^{h} = \partial_r \Gamma_{s\,i}^{\,h} - \partial_s \Gamma_{r\,i}^{\,h}.$$

Therefore

$$\nabla_r(\nabla_s v^h) - \nabla_s(\nabla_r v^h) = R_{i\,,\,rs}^{h} v^i. \tag{84.1}$$

Since the terms of this identity are tensors, (84.1) is valid for any system of coordinates and for any point of V_n. If it had been established directly, the identity (84.1) might have been used to introduce the Riemann-Christoffel tensor. This tensor was, in fact, originally introduced in this way.

85. The Ricci tensor. Let us investigate the tensors which can be formed from the Riemann-Christoffel tensor by contraction. Since the components $R_{ij,\,rs}$ are antisymmetric on the one hand with respect to i and j and on the other with respect to r and s, contraction of i with j, or of r with s merely results in a tensor which is identically zero.

Consider, however, the contraction of an index from the first group with one from the second – using, for example, the second and third indices. In this way we obtain the tensor

$$R_{ij} = R_i{}^h{}_{,hj}. \qquad (85.1)$$

This tensor is evidently *symmetric* with respect to the indices i and j, since

$$R_{ij} = g^{hk} R_{ik,\,hj} = g^{hk} R_{hj,\,ik} = g^{hk} R_{jh,\,ki} = R_{ji}.$$

If we contract either index of the first group with either one of the second we always obtain either the tensor R_{ij} or its negative since, from (83.3)

$$g^{hk} R_{ki,\,jh} = g^{hk} R_{ik,\,hj} = R_{ij}$$

and $\qquad g^{hk} R_{ik,\,jh} = g^{hk} R_{ki,\,hj} = -g^{hk} R_{ik,\,hj} = -R_{ij}.$

The symmetric tensor R_{ij}, which plays a fundamental role in relativistic gravitational theory, is known as the Ricci tensor. Using (81.2) and (85.1) we can deduce the following expression for this tensor:

$$R_{ij} = \partial_h \Gamma_i{}^h{}_j - \partial_j \Gamma_h{}^h{}_i + \Gamma_h{}^h{}_l \Gamma_i{}^l{}_j - \Gamma_i{}^l{}_h \Gamma_j{}^h{}_l. \qquad (85.2)$$

Contracting the Ricci tensor we obtain the invariant

$$R = R_i{}^i = g^{ij} R_{ij} \qquad (85.3)$$

which is called the *scalar Riemannian curvature* of the space V_n.

In the Riemannian space realized by an ordinary two-dimensional locally deformed surface, the scalar Riemannian curvature is equivalent to what is known, in differential geometry, as the total curvature of the surface. This total curvature is known to depend only upon ds^2, the line element in the surface.

86. Bianchi identities. There exist identities other than those implied by (83.3) and (83.4), between the components $\nabla_t R_{ij,\,rs}$ of the tensor derivative of the curvature tensor.

In order to establish these important identities we again use a system of normal coordinates at an arbitrary point M of V_n. The Christoffel symbols are then zero at M and using (81.2) we deduce by differentiation that, at the point M,

$$\nabla_t R_i{}^h{}_{,rs} = \partial_{rt}\Gamma_s{}^h{}_i - \partial_{st}\Gamma_r{}^h{}_i. \tag{86.1}$$

A cyclic permutation of the indices r, s and t gives

$$\nabla_r R_i{}^h{}_{,st} = \partial_{sr}\Gamma_t{}^h{}_i - \partial_{tr}\Gamma_s{}^h{}_i,$$

$$\nabla_s R_i{}^h{}_{,tr} = \partial_{ts}\Gamma_r{}^h{}_i - \partial_{rs}\Gamma_t{}^h{}_i.$$

Adding these expressions we obtain

$$\nabla_r R_i{}^h{}_{,st} + \nabla_s R_i{}^h{}_{,tr} + \nabla_t R_i{}^h{}_{,rs} = 0. \tag{86.2}$$

From their tensorial form these identities are obviously valid in any coordinate system and for any point of V_n. They are known as the *Bianchi identities*.

Double contraction of (86.2) leads to an important consequence relating to the Ricci tensor. Putting $t = h$ we obtain

$$-\nabla_r R_{is} + \nabla_s R_{ir} + \nabla_h R_i{}^h{}_{,rs} = 0,$$

then, contracting i and s,

$$-\nabla_r R + \nabla_s R^s{}_r + \nabla_h R^h{}_r = 0,$$

or

$$2\nabla_s R^s{}_r - \nabla_r R = 0.$$

Using Ricci's theorem this may be rewritten in the form

$$\nabla_s(R^s{}_r - \tfrac{1}{2}g^s{}_r R) = 0. \tag{86.3}$$

It follows that the symmetric tensor

$$S_{rs} = R_{rs} - \tfrac{1}{2}g_{rs}R \tag{86.4}$$

satisfies the identities

$$\nabla_s S^s{}_r = 0. \tag{86.5}$$

These identities are fundamental to relativistic gravitation theory in which they are used to introduce conservation principles.

PART II: APPLICATIONS

CHAPTER VI

Tensor Calculus and Classical Dynamics

I. DYNAMICS OF HOLONOMIC SYSTEMS WITH TIME-INDEPENDENT CONSTRAINTS

87. Configuration space as a Riemannian space. Consider a dynamical system S with time independent holonomic constraints and n degrees of freedom. The set of configurations of such a system constitutes an n-dimensional differentiable continuum which is called *configuration space*. We describe S by the parameters $(q^1, \ldots, q^n)$ which form a coordinate system in the configuration space. According to our definition of S, its kinetic energy T is a positive definite quadratic form in the time derivatives of the q^i;

$$2T = a_{ij} q'^i q'^j \quad \left(q'^i \equiv \frac{dq^i}{dt} \right), \tag{87.1}$$

where the a_{ij} are functions of the parameters q^i. We can associate the dynamical system S with the proper Riemannian space V_n defined by the configuration space associated with the metric

$$ds^2 = 2T \, dt^2,$$

which may be rewritten

$$ds^2 = a_{ij} dq^i dq^j. \tag{87.2}$$

A definite point M of the configuration space corresponds to each configuration of the system in such a way that a displacement of the point M in the Riemannian space V_n is associated with every displacement of the system. We propose to represent the motion of the system S by that of a point in the Riemannian space.

[98]

88. Kinematics of the point M. Let us first complete the kinematic considerations sketched in Chapter V. The velocity vector, **v**, of the point M whose coordinates are q^i, has the contravariant components

$$v^i = \frac{dq^i}{dt} = q'^i.$$

The covariant components of this vector are consequently given by

$$v_i = a_{ij}q'^j = \frac{\partial T}{\partial q'^i}. \tag{88.1}$$

We thus see that the *momenta* p_i, which continually appear in analytical dynamics, are just the covariant components of the velocity vector of the representative point M in the Riemannian space V_n. Let us denote the unit vector tangential to the trajectory, C, of M by **u**; this has the components

$$u^i = \frac{dq^i}{ds}.$$

Hence

$$\mathbf{v} = v\mathbf{u}, \qquad v = \frac{ds}{dt},$$

and the magnitude of the velocity vector is given by

$$v^2 = \left(\frac{ds}{dt}\right)^2 = 2T. \tag{88.2}$$

Now consider the acceleration vector. Since **u** is a unit vector its absolute differential, as M describes C, is perpendicular to it. Consequently we may put

$$\frac{\nabla u^i}{ds} = \frac{n^i}{\rho}, \tag{88.3}$$

where the n^i denote the components of a unit vector normal to **u** and ρ is a positive scalar quantity. The vector **n** will be called the principal normal vector to C and ρ^{-1} the curvature of C in V_n. (88.3) is a generalization of Frenet's first formula in elementary differential geometry. Using

$$v^i = vu^i,$$

and differentiating with respect to time, we have

$$\gamma^i = \frac{\nabla(vu^i)}{dt} = \frac{dv}{dt}u^i + v\frac{\nabla u^i}{ds}\frac{ds}{dt},$$

or
$$\gamma^i = \frac{dv}{dt}u^i + \frac{v^2}{\rho}n^i. \tag{88.4}$$

Hence the acceleration vector γ decomposes into a tangential and a normal acceleration given by the same expressions as in classical mechanics.

89. Equations of motion. Let

$$Q_i\,dq^i$$

denote the infinitesimal amount of work done by the given applied forces acting on S in a small arbitrary virtual displacement. This expression is invariant under a change of parameters if the Q_i are the covariant components of a vector in V_n. The vector Q_i is called the *generalized force vector*. The motion of S is determined by the Lagrange equations†

$$P_i \equiv \frac{d}{dt}\left(\frac{\partial T}{\partial q'^i}\right) - \frac{\partial T}{\partial q^i} = Q_i. \tag{89.1}$$

Let us try to interpret the components of P_i. We have

$$\frac{\partial T}{\partial q'^i} = a_{ij}q'^j, \qquad \frac{\partial T}{\partial q^i} = \tfrac{1}{2}\partial_i a_{jk}q'^j q'^k.$$

Hence the P_i have the explicit form

$$P_i = \frac{d}{dt}(a_{ij}q'^j) - \tfrac{1}{2}\partial_i a_{jk}q'^j q'^k. \tag{89.2}$$

† These equations are discussed in any elementary account of analytical mechanics. They are derived from Newton's laws of motion which relate the forces to the acceleration of the individual particles of the system. (T.)

However, from a calculation identical with that given in §76, we see that the right-hand side of (89.2) is equal to

$$a_{ij}\frac{dq'^j}{dt} + [jk, i]q'^j q'^k,$$

where the Christoffel symbols are defined in terms of the a_{ij}. We deduce that

$$P_i = a_{ij}\frac{\nabla v^j}{dt} = \frac{\nabla v_i}{dt} = \gamma_i. \tag{89.3}$$

Thus the quantities P_i, which appear on the left-hand side of the Lagrange equations, are just the covariant components of the acceleration vector of M, and the Lagrange equations may be rewritten as

$$\gamma_i = Q_i. \tag{89.4}$$

The equations of motion of M are thus obtained by equating the acceleration vector in V_n to the generalized force vector; that is by writing down for M the precise generalization of the equation of motion in particle dynamics and considering M to have unit mass.

Using (88.4) we may write the equations of motion as

$$\frac{dv}{dt}u_i + \frac{v^2}{\rho}n_i = Q_i. \tag{89.5}$$

It follows that the generalized force vector is always coplanar with the tangent and the principal normal of the trajectory of M.

If the Q_i are zero, the point M has zero acceleration in V_n, which implies that

$$\frac{dv}{dt} = 0, \quad \frac{1}{\rho} = 0.$$

The motion of S in the absence of forces is therefore associated with the uniform motion of M along geodesics in V_n.

90. Energy integral. Suppose that the system S can be associated with a *time independent potential function U of the parameters* q^i which satisfies

$$Q_i = -\partial_i U.$$

Then taking the scalar product of each side of (89.5) with the vector **v** which has the components $v^i = vu^i$ we find, allowing for the orthogonality of **v** and **n**,

$$v\frac{dv}{dt} = -\partial_i U \frac{dq^i}{dt} = -\frac{dU}{dt}.$$

It follows that the equations of motion of M have a first integral which is a generalization of the energy integral relating to a single particle in elementary dynamics:

$$\tfrac{1}{2}v^2 + U = h \quad (h = \text{const.}). \tag{90.1}$$

According to (88.2) this is also the energy integral of the physical system S.

91. Principle of Maupertuis. We retain the assumption which led to an energy integral in the previous section and take a particular value for the arbitrary constant appearing in the function U. It is then easy to derive a generalization of the principle of Maupertuis regarding motion in the absence of forces and to give a geometrical interpretation of this.

Consider a parametric representation of the motion of M in which the coordinates q^i and the time t are expressed as functions of some parameter τ. We suppose that the relation between τ and t is known, and write

$$\frac{d\tau}{dt} = F. \tag{91.1}$$

We have

$$q'^i = \frac{dq^i}{dt} = F\frac{dq^i}{d\tau},$$

and the quantities P_i may be written

$$P_i = F\left[\frac{d}{d\tau}\left(Fa_{ij}\frac{dq^j}{d\tau}\right) - \frac{F}{2}\partial_i a_{jk}\frac{dq^j}{d\tau}\frac{dq^k}{d\tau}\right].$$

Hence we deduce that the trajectory, C, of M can be represented in terms of the parameter τ as the solution of the system of differential equations

$$\frac{d}{d\tau}\left(Fa_{ij}\frac{dq^j}{d\tau}\right) - \frac{F}{2}\partial_i a_{jk}\frac{dq^j}{d\tau}\frac{dq^k}{d\tau} = -\frac{\partial_i U}{F}. \tag{91.2}$$

The form of the left-hand side of (91.2) brings to mind the equations which determine the extremals of

$$\sigma = \int \sqrt{(Fa_{ij}dq^i dq^j)}, \tag{91.3}$$

where F is a function of the q^i. Let us determine the extremals of (91.3) assuming F to be a known function. These correspond to the geodesics of the Riemannian metric

$$d\sigma^2 = F ds^2 = Fa_{ij}dq^i dq^j. \tag{91.4}$$

We take σ itself as parameter on these geodesics and, in order to simplify the notation, we put

$$\dot{q}^i = \frac{dq^i}{d\sigma}.$$

It follows from (76.7) that the differential equations which determine the geodesics can be written

$$\frac{d}{d\sigma}(Fa_{ij}\dot{q}^j) - \tfrac{1}{2}\partial_i(Fa_{jk})\dot{q}^j\dot{q}^k = 0,$$

so that

$$\frac{d}{d\sigma}(Fa_{ij}\dot{q}^j) - \frac{F}{2}(\partial_i a_{jk})\dot{q}^j\dot{q}^k = \tfrac{1}{2}(\partial_i F)a_{jk}\dot{q}^j\dot{q}^k. \tag{91.5}$$

From (91.4)

$$Fa_{jk}\dot{q}^j\dot{q}^k = 1,$$

so that (91.5) may be rewritten

$$\frac{d}{d\sigma}(Fa_{ij}\dot{q}^j) - \frac{F}{2}\partial_i a_{jk}\dot{q}^j\dot{q}^k = \frac{1}{2}\frac{\partial_i F}{F}. \tag{91.6}$$

We can identify (91.2) with (91.6) if we take

$$F = -2U + \text{const.},$$

so that

$$\partial_i F = -2\partial_i U.$$

Among all the trajectories which are solutions of (91.2) consider those which correspond to a given value of the energy constant h. (88.2) substituted into (90.1) gives

$$2T = 2(h - U). \tag{91.7}$$

The parameter τ is completely determined on these trajectories (apart from an additive constant) if we put

$$F = 2(h - U). \tag{91.8}$$

Using (91.4), the parameter σ is determined by

$$\frac{d\sigma}{dt} = \sqrt{(F.2T)}$$

or, from (91.7),

$$\frac{d\sigma}{dt} = F \equiv \frac{d\tau}{dt}.$$

The parameters σ and τ are therefore identical and the trajectories under consideration satisfy the differential system (91.6), that is to say they are the geodesics of the Riemannian metric

$$d\sigma^2 = 2(h - U) a_{ij} dq^i dq^j.$$

We now express this formally:

MAUPERTUIS' PRINCIPLE: *Given the form of the energy integral, the trajectories of a dynamical system which are determined by a particular value, h, of the energy constant are the geodesics of the configuration space with the Riemannian metric*

$$d\sigma^2 = 2(h - U) ds^2 = 2(h - U) a_{ij} dq^i dq^j. \tag{91.9}$$

These geodesics are given by

$$\frac{d\sigma}{dt} = 2(h - U). \tag{91.10}$$

92. Some applications. The introduction of Riemannian spaces associated with the configuration spaces gives us a geometrical picture for all problems concerning the dynamics of holonomic systems with time independent constraints. The techniques of tensor calculus may therefore be applied to such problems. The use of tensor methods in analytical dynamics has resolved a certain number of problems for which solutions could not be obtained using only the methods of analytical dynamics. This is so, in particular, in the case of the problem of the *transformation of the equations of motion* which was posed by Painlevé and Levi-Civita and which can be formulated as follows: Under what conditions do two systems S and S', which share the same configuration space, have the same trajectories independently of the way in which these trajectories are described in time ?†

The *stability* of the motion of a dynamical system S is best discussed by reference to the configuration space V_n with the metric $ds^2 = 2T dt^2$ using a method analogous to Levi-Civita's method of geodesic spread.‡

We mention finally the research on *reducible dynamical systems* §: a dynamical system is said to be reducible if, for a certain choice of parameters q^i, its energy T can be expressed as the sum of the energies of two systems having respectively r and $(n-r)$ degrees of freedom

$$2T_1 = \sum_1^r a_{\alpha\beta}(q^1,\ldots,q^r)q'^\alpha q'^\beta,$$

$$2T_2 = \sum_{r+1}^n b_{\alpha\beta}(q^{r+1},\ldots,q^n)q'^\alpha q'^\beta,$$

the potential function U also being the sum of two potential functions

$$U_1(q^1,\ldots,q^r), \qquad U_2(q^{r+1},\ldots,q^n).$$

† PAINLEVÉ, 'Sur la transformation des équations de la dynamique', *Journ. de Math.* (1894); LEVI-CIVITA, 'Sulle transformazioni delle equazioni dinamiche', *Ann. di Mat.* (1896); THOMAS, T. Y., 'On the transformation of the equations of dynamics', *Journ. of Math. Phys.* (1946); LICHNEROWICZ, 'Sur la transformation des équations de la dynamique', *C. R. Acad. Sci.* (1946).

‡ LEVI-CIVITA, 'Sur l'écart géodésique', *Math. Ann.* (1926); SYNGE, 'On the geometry of dynamics', *Trans. Roy. Soc., Lond.* (1926).

§ STACKEL, *C. R. Acad. Sci.* (1895); THOMAS, T. Y., 'Reducible dynamical systems', *Journ. of Math. Phys.* (1947).

Given a dynamical system S referred to any set of parameters $(q^{i'})$, it is of interest to determine under what conditions it is possible to find a set of parameters (q^i) which allow the reduction of the system.

The explicit solution of these problems goes beyond the scope of this book and we refer the interested reader to the original papers. It is worth noting that, to some extent, the considerations of previous sections can be extended to non-holonomic systems with time independent constraints.†

II. DYNAMICS OF HOLONOMIC SYSTEMS WITH TIME-DEPENDENT CONSTRAINTS

93. Configuration space-time. Consider a dynamical system Σ having n degrees of freedom and with time-dependent holonomic constraints. Since the constraints are time dependent the possible configurations of the system depend upon the instant of time under consideration. We therefore substitute a configuration space-time for configuration space, that is a continuum V_{n+1} with $(n+1)$ dimensions which has the parameters q^i $(i = 1,\ldots,n)$ of the system and the time $t = q^0$, as coordinates. This coordinate system is restricted by allowing only coordinate transformations of the type

$$q^{i'} = f^{i'}(q^0,q^i) \qquad q^{0'} = q^0 + \phi(q^i)$$

$$(i, i', \text{etc.} = 1,\ldots,n).$$

It will be convenient to write

$$q'^{\alpha} = \frac{dq^{\alpha}}{dt} \qquad (\alpha, \text{etc.} = 0, 1,\ldots,n)$$

so that

$$q'^0 = 1, \qquad q'^i = \frac{dq^i}{dt}.$$

In this notation the energy of Σ is given by

$$2T = a_{\alpha\beta}q'^{\alpha}q'^{\beta}, \tag{93.1}$$

† See, for example, the work by Synge just quoted.

where the $a_{\alpha\beta}$ are functions of the q^{λ}. We associate the dynamical system Σ with the Riemannian space V_{n+1} defined by configuration space-time with the metric

$$ds^2 = 2T \, dt^2 = a_{\alpha\beta} \, dq^{\alpha} \, dq^{\beta}. \tag{93.2}$$

Each configuration of the system at time t corresponds to a completely determined point M of configuration space-time. Each possible displacement of the system is associated with the displacement of the point M in the Riemannian space V_{n+1}, this displacement being defined by giving the n coordinates q^i as functions of the $(n+1)$-th coordinate $q^0 = t$.

The velocity vector of M has the contravariant components $v^{\alpha} = q'^{\alpha}$ and its covariant components are given by

$$v_{\alpha} = a_{\alpha\beta} q'^{\beta} = \frac{\partial T}{\partial q'^{\alpha}}.$$

The n components v_i are thus equal to the momentum components, p_i, of the system Σ. The acceleration vector γ has the covariant components

$$\gamma_{\alpha} = a_{\alpha\beta} \frac{\nabla v^{\beta}}{dt} = a_{\alpha\beta} q''^{\beta} + [\beta\gamma, \alpha] q'^{\beta} q'^{\gamma}. \tag{93.3}$$

We note that $q''^0 = 0$. In particular it follows from (93.3) that

$$\gamma_0 = a_{0i} q''^{i} + [jk, 0] q'^{j} q'^{k} + \partial_i a_{00} q'^{i} + \tfrac{1}{2} \partial_0 a_{00}. \tag{93.4}$$

94. Equations of motion. Once again let

$$Q_i \, dq^i$$

be the infinitesimal amount of work done in a small arbitrary virtual displacement by the forces applied to Σ. The motion of Σ is determined by the Lagrange equations

$$P_i \equiv \frac{d}{dt}\left(\frac{\partial T}{\partial q'^{i}}\right) - \frac{\partial T}{\partial q^i} = Q_i. \tag{94.1}$$

Multiplying these by q'^i and summing

$$\frac{d}{dt}\left(q'^i \frac{\partial T}{\partial q'^i}\right) - \left(\frac{\partial T}{\partial q'^i} q''^i + \frac{\partial T}{\partial q^i} q'^i\right) = Q_i q'^i.$$

Using Euler's theorem for homogeneous functions, we have

$$q'^i \frac{\partial T}{\partial q'^i} + q'^0 \frac{\partial T}{\partial q'^0} = 2T$$

also

$$q''^i \frac{\partial T}{\partial q'^i} + q'^i \frac{\partial T}{\partial q^i} = \frac{dT}{dt} - \frac{\partial T}{\partial q^0},$$

so that

$$P_0 \equiv \frac{d}{dt}\left(\frac{\partial T}{\partial q'^0}\right) - \frac{\partial T}{\partial q^0} = \frac{dT}{dt} - Q_i q'^i. \tag{94.2}$$

We now wish to interpret the quantities P_α which appear in (94.1) and (94.2). As in §**89** we have

$$P_\alpha = a_{\alpha\beta} q''^\beta + [\beta\gamma, \alpha] q'^\beta q'^\gamma.$$

Putting

$$P_\alpha = \gamma_\alpha,$$

the Lagrange equations (94.1) become

$$\gamma_i = Q_i \tag{94.3}$$

and (94.2) takes the form

$$\gamma_0 = \frac{dT}{dt} - Q_i q'^i. \tag{94.4}$$

Equations (94.3) and (94.4) are the *equations of motion* of M in V_{n+1}. If the motion of S takes place in the absence of applied forces the n components γ_i of the acceleration of S are zero, but γ_0 is, in general, different from zero and the corresponding trajectories of M in V_{n+1} cannot be interpreted in a simple geometrical manner.

95. Systems with a potential function. Suppose that the forces acting on Σ are derivable from a potential function $U(q^0, q^1, \ldots, q^n)$ which is explicitly time dependent. We know that if we introduce the Lagrangian of the system

$$L = T - U,$$

then the equations of motion of Σ can be written

$$\frac{d}{dt}\left(\frac{\partial L}{\partial q'^i}\right)-\frac{\partial L}{\partial q^i} = 0.$$

Instead of $ds^2 = 2T\,dt^2$, it is now convenient to introduce the metric

$$d\sigma^2 = 2L\,dt^2. \tag{95.1}$$

Absorbing the function U in the coefficient a_{00} we write

$$2L = a_{\alpha\beta}q'^\alpha q'^\beta,$$

so that $\qquad\qquad d\sigma^2 = a_{\alpha\beta}dq^\alpha dq^\beta.$

All the formulae established in §93 and §94 remain valid in terms of the metric (95.1) if we replace T by L and write $Q_i = 0$. We thus have

$$\gamma_\alpha = \frac{d}{dt}\left(\frac{\partial L}{\partial q'^\alpha}\right)-\frac{\partial L}{\partial q^\alpha},$$

and the equations of motion of the representative point M in the space V_{n+1} with the metric (95.1) can be written

$$\gamma_i = 0, \qquad \gamma_0 = \frac{dL}{dt}. \tag{95.2}$$

96. Eisenhart's theorem. Eisenhart has given a simple geometrical interpretation of (95.2) by introducing a Riemannian metric with $(n+2)$ dimensions. If $u = q^{n+1}$ denotes a supplementary parameter, consider the improper Riemannian metric defined by

$$d\tau^2 = 2L\,dt^2 + 2dt\,du, \tag{96.1}$$

giving

$$d\tau^2 = a_{ij}dq^i dq^j + 2a_{i0}dq^i dq^0 + a_{00}(dq^0)^2 + 2dq^0 dq^{n+1}. \tag{96.2}$$

Among the geodesics of such a metric there are some which are real and of zero length, that is with $d\tau^2 = 0$. For simplicity we shall consider only these geodesics in what follows. If we write the metric (96.2) in the condensed form

$$d\tau^2 = a_{AB}dq^A dq^B \quad (A, B, \text{etc.} = 0, 1, \ldots, n+1)$$

with

$$a_{0\overline{n+1}} = 1, \qquad a_{i\overline{n+1}} = a_{\overline{n+1}\,\overline{n+1}} = 0,$$

then the system of differential equations (96.2) which determine the geodesics may be written

$$a_{AB}\frac{d^2q^B}{d\tau^2} + [BC, A]\frac{dq^B}{d\tau}\frac{dq^C}{d\tau} = 0. \qquad (96.3)$$

When $A = n+1$, we have simply

$$\frac{d^2t}{d\tau^2} = 0, \qquad \frac{dt}{d\tau} = a \quad (a = \text{const.}). \qquad (96.4)$$

Using (96.1) and (96.4) it follows that

$$\frac{du}{dt} = \frac{1}{2}\left(\frac{d\tau}{dt}\right)^2 - L = \frac{1}{2a^2} - L \qquad (96.5)$$

and, by integration

$$u = \frac{t}{2a^2} - \int_0^t L\,dt + b \quad (b = \text{const.}). \qquad (96.6)$$

The differential system (96.3) may then be written

$$a_{AB}\frac{d^2q^B}{dt^2} + [BC, A]\frac{dq^B}{dt}\frac{dq^C}{dt} = 0.$$

When $A = i$ we get

$$\gamma_i = 0$$

and, for $A = 0$,

$$\frac{d^2u}{dt^2} + \gamma_0 = 0$$

or, using (96.5),

$$\gamma_0 = \frac{dL}{dt},$$

and we recover (95.2) which are the equations of motion of the system Σ. A consideration of the initial conditions enables us to deduce the following theorem† :

† EISENHART, 'Dynamical trajectories and geodesics', *Ann. of Math.* (1929).

EISENHART'S THEOREM: *All possible motions of a holonomic system Σ which admits a potential function can be obtained in the following way: consider the metric (96.1); the possible motions of Σ correspond to the geodesics of this metric which satisfy the initial conditions*

$$u_0 = b, \qquad \left(\frac{du}{dt}\right)_0 = \frac{1}{2a^2} - (L)_0,$$

where a and b are arbitrary constants.

The trajectories of the representative point M of V_{n+1} are given in configuration space-time by the projections of the geodesics of (96.1) along the coordinate lines of the parameter u. The relation (96.6) shows the connection between the variable u and the action of the system. We remark finally that, given Eisenhart's theorem, the Hamilton-Jacobi theorem of analytical mechanics can be interpreted in a particularly simple manner.

III. DYNAMICS OF CONTINUOUS MEDIA

97. Continuous media. In a later part of this chapter we propose to indicate some applications of tensor calculus to the dynamics of continuous media. This includes the hydrodynamics of fluids as well as the study of the deformation of solids (theory of elasticity). Historically, it was the work of the physicist Voigt on the deformation of crystalline media that gave rise to the idea of tensors and it is to the theory of elasticity that they owe their name. This, of all the classical theories, is probably the one where tensor methods have proved most useful, thanks to the easy introduction of those curvilinear coordinates which are most appropriate to the physical problems under consideration.† Lastly, the general equations for continuous media which we shall establish are very important because of Einstein's use of them in general relativity theory.

From the microscopic point of view all physical media consist of particles. It is possible, however, to adopt a macroscopic viewpoint and to describe the behaviour of a continuous medium such as a

† See BRILLOUIN, L., *Les tenseurs en mécanique et en élasticité* (Masson, 1938), on this point.

fluid or elastic solid by considering a small volume of the medium rather than the individual particles. In the course of time, particles enter and leave such a volume element, each particle obeying the usual laws of mechanics. It is necessary to reformulate these laws in such a way that the positions and velocities of the individual particles do not appear. The volume elements considered must therefore contain a sufficiently large number of particles in order that the mean values of these quantities should be well defined. If the space is referred to three arbitrary curvilinear coordinates y^1, y^2, y^3 we assume that it is possible to define a density of matter ρ and a vector velocity **v** at the point $M(y^i)$ and at the instant t. The variables (y^1, y^2, y^3, t) are known as the *Euler variables*.

98. Total and partial time derivatives. Let us consider a function of position, q, in a continuous medium (for example, a scalar such as ρ, or a component of a vector such as **v**, etc.) and suppose that we wish to describe its evolution in time. This can be done in two ways: either we describe its evolution in time at a fixed point $M(y^i)$ so that the derivative of q with respect to time is the partial derivative given by

$$\frac{\partial q}{\partial t} = \left(\frac{\partial q}{\partial t}\right)_{y^i = \text{const.}}, \tag{98.1}$$

or we can refer this evolution to a coordinate system which is tied locally to the average motion of the matter. In the latter case the time derivative of q is the total derivative

$$\frac{dq}{dt} = \frac{\partial q}{\partial t} + \frac{\partial q}{\partial y^1}\frac{dy^1}{dt} + \frac{\partial q}{\partial y^2}\frac{dy^2}{dt} + \frac{\partial q}{\partial y^3}\frac{dy^3}{dt}.$$

Introducing the contravariant components of **v** defined by

$$v^i = \frac{dy^i}{dt},$$

and using the usual notation for partial derivatives, this relation becomes

$$\frac{dq}{dt} = \frac{\partial q}{\partial t} + v^i \partial_i q. \tag{98.2}$$

If q is a scalar (98.2) can be written in vectorial form as

$$\frac{dq}{dt} = \frac{\partial q}{\partial t} + \mathbf{v} \cdot \operatorname{grad} q. \tag{98.3}$$

99. The continuity equation. The two purely kinetic quantities, ρ and $\mathbf{v}$, are not independent since the change of density in a volume element is determined by the flux of matter which crosses the surface of this element. This is expressed mathematically by the well-known *continuity equation*

$$\frac{\partial \rho}{\partial t} + \operatorname{div}(\rho \mathbf{v}) = 0, \tag{99.1}$$

which may be expressed in tensor form as

$$\frac{\partial \rho}{\partial t} + \nabla_i(\rho v^i) = 0, \tag{99.2}$$

where the covariant differentiation is performed in a Euclidean space of three dimensions referred to the curvilinear coordinates y^i.

100. Body and surface forces. In order to deal with the dynamics of continuous media it is necessary to investigate the forces which are exerted on the volume elements of the medium. Draw a closed surface S inside the medium and consider the external forces acting on the volume, V, of the medium enclosed by S. These forces fall naturally into two classes:

(*1*) *Body forces*: these act directly on the different volume elements of V. Choose an element dV of V with mass $dm = \rho\, dV$. The forces acting on this volume element are of the order of dm and hence dV; we denote their resultant by

$$\mathbf{f}\, dV,$$

where $\mathbf{f}$, in general a function of position, is the body force per unit volume.

(*2*) *Surface forces*: these are the forces acting on the surface which arise from the action between the elements of the medium contiguous to S on the outside and on the inside. If dS is a surface element then it is acted upon by a force of order dS which we denote by

$$-\mathbf{T}\, dS.$$

This force depends only upon the position of the element dS in the medium and not upon the rest of the surface which contains it. It is generally inclined at an angle to the normal of the surface element dS; when directed towards the interior of S it is called a *compression* and in the reverse case, a *tension*.

101. The stress tensor. We now propose to investigate the variation of the force $\mathbf{T} \, dS$ with the different possible orientations of dS.

To this end we refer the continuous medium to a frame (O, x^1, x^2, x^3) formed by three rectangular Cartesian coordinate axes. At any point M of the medium consider three lines parallel to the axes and take three infinitesimal elements along these. In this way we obtain an elementary tetrahedron $MABC$ whose faces MBC, MCA, MAB, ABC will be denoted respectively by dS^{23}, dS^{31}, dS^{12} and dS. If α_1, α_2, α_3 denote the components of the unit vector along the normal to dS exterior to the tetrahedron we have

$$dS^{23} = \alpha_1 \, dS, \qquad dS^{31} = \alpha_2 \, dS, \qquad dS^{12} = \alpha_3 \, dS.$$

The surface force acting on ABC differs only by higher order terms from that which would be exerted on a similar parallel surface centred at M. Consequently we denote by $\mathbf{T} \, dS$ the surface force exerted over dS by the interior elements of the tetrahedron. The equality of action and reaction implies that the force exerted over dS by the external elements will be $-\mathbf{T} \, dS$.

The tetrahedron is therefore subjected to the body force $\mathbf{f} \, dV$ and to the surface forces which act on its four faces and may be represented by

$$\Theta_{23} \, dS^{23}, \quad \Theta_{31} \, dS^{31}, \quad \Theta_{12} \, dS^{12}, \quad -\mathbf{T} \, dS.$$

If we add to these the inertial force $-\gamma \, dm$ corresponding to the acceleration of the point M of the medium, the elementary tetrahedron will be in equilibrium under the action of the complete system of forces. Hence

$$-\mathbf{T} \, dS + \Theta_{23} \, dS^{23} + \Theta_{31} \, dS^{31} + \Theta_{12} \, dS^{12} + (\mathbf{f} - \rho\gamma) \, dV = 0,$$

and dividing by dS,

$$\mathbf{T} = \alpha_1 \Theta_{23} + \alpha_2 \Theta_{31} + \alpha_3 \Theta_{12} + (\mathbf{f} - \rho\gamma) \frac{dV}{dS}.$$

Suppose that the lengths of the edges of the tetrahedron tend to zero, then dV/dS tends to zero, and in the limit we have

$$\mathbf{T} = \alpha_1 \mathbf{\Theta}_{23} + \alpha_2 \mathbf{\Theta}_{31} + \alpha_3 \mathbf{\Theta}_{12}. \qquad (101.1)$$

We see that, if dS^{23}, dS^{31}, dS^{12} represent the components of the bivector† defining the area dS, then

$$\mathbf{T} \, dS = \mathbf{\Theta}_{23} \, dS^{23} + \mathbf{\Theta}_{31} \, dS^{31} + \mathbf{\Theta}_{12} \, dS^{12}.$$

Hence, in our coordinate system x^i, the components of the force $\mathbf{T} \, dS$ are linear functions of the components of the bivector defining the surface element dS. This result is obviously independent of the particular coordinate system used.

Let us now refer the space to an arbitrary system of curvilinear coordinates y^i. The contravariant components of the bivector defining the element dS are denoted by dS^{ij} and the contravariant components of the force $\mathbf{T} \, dS$ are denoted by $T^l \, dS$. According to the above result there exists a set of quantities Θ^l_{ij} such that

$$T^l \, dS = \tfrac{1}{2} \Theta^l_{ij} \, dS^{ij}.$$

In place of the bivector dS^{ij} it is convenient to introduce the adjoint‡ vector whose covariant components $d\sigma_k$ are defined by

$$dS^{ij} = \epsilon^{kij} \, d\sigma_k. \qquad (101.2)$$

It can immediately be verified that $(d\sigma_k)$ is orthogonal to the surface element dS and has the same magnitude. It follows that

$$T^l \, dS = \tfrac{1}{2} \Theta^l_{ij} \eta^{kij} \, d\sigma_k,$$

and writing $\qquad\qquad t^{kl} = \tfrac{1}{2} \Theta^l_{ij} \eta^{kij},$

we obtain the fundamental relations

$$T^l \, dS = t^{kl} \, d\sigma_k. \qquad (101.3)$$

It is clear that the quantities t^{kl} are the contravariant components of a tensor since, whatever the numerical values of the covariant components $d\sigma_k$, the left-hand side of (101.3) is a contravariant vector.

† Defined in §47. (T.) ‡ Defined in §52.

The tensor t^{kl} so defined† is known as the *stress tensor* of the continuous medium.

When the medium is a perfect fluid the stress tensor takes the form

$$t^{kl} = pg^{kl} \qquad (101.4)$$

where p is the scalar pressure in the fluid at the point and time instant considered and where the g^{kl} are the components of the fundamental tensor of the space.

102. General dynamical equations for continuous media. In the following we shall assume that all the quantities introduced are both continuous and differentiable.

Consider again an arbitrary closed surface S within the medium containing a region of volume V and write down the usual conditions for the resultant of the exterior forces and the forces of inertia acting on this region of the medium to vanish. It is convenient, particularly for writing down the moment conditions, to use an orthogonal rectilinear system of coordinates x^i.

Let γ^i be the components of the acceleration of the centre of mass $M(x^i)$ of an element of volume of V; the corresponding inertial force acting on the element has components $-\rho\gamma^i dV$. If $\mathbf{T} dS$ represents the surface force exerted by the elements interior to S on the elements exterior to S it follows from the vanishing of the resultant force on V that

$$\iiint_V (f^i - \rho\gamma^i)\,dV - \iint_S t^{ki}\,d\sigma_k = 0 \qquad (102.1)$$

and, since the moment of the resultant couple also vanishes,

$$\iiint_V [x^i(f^j - \rho\gamma^j) - x^j(f^i - \rho\gamma^i)]\,dV -$$

$$- \iint_S (x^i t^{kj} - x^j t^{ki})\,d\sigma_k = 0. \quad (102.2)$$

Using Green's theorem to transform the surface integral appearing in (102.1) we obtain

$$\iiint_V (f^i - \rho\gamma^i - \partial_k t^{ki})\,dV = 0 \qquad (102.3)$$

† Note that in many English texts the stress tensor is defined with the opposite sign. (T.)

and (102.2) becomes

$$\iiint_V [x^i(f^j - \rho\gamma^j - \partial_k t^{kj}) - x^j(f^i - \rho\gamma^i - \partial_k t^{ki})]\,dV -$$

$$- \iiint_V (t^{ij} - t^{ji})\,dV = 0. \quad (102.4)$$

Since (102.3) holds for any volume V it follows from the continuity of the integrand that this is identically zero, hence

$$\rho\gamma^i = f^i - \partial_k t^{ki}. \quad (102.5)$$

Using this relation (102.4) reduces to

$$\iiint_V (t^{ij} - t^{ji})\,dV = 0,$$

for an arbitrary volume V so that

$$t^{ij} - t^{ji} = 0. \quad (102.6)$$

Equations (102.5) and (102.6) have been derived from the laws of classical mechanics. The first set comprise the general dynamical relations for continuous media, the second merely state that the tensor t^{ij} is symmetric with respect to its two indices.

It is easy to express the dynamical equations for continuous media using an *arbitrary curvilinear coordinate system*. The equations

$$\rho\gamma^i = f^i - \nabla_k t^{ki} \quad (102.7)$$

are invariant with respect to any change of coordinate system; they also reduce to the equations obtained previously when using rectangular axes. They are therefore the desired generalization of (102.5). We note that to any stress tensor t^{ij} there corresponds a force density per unit volume given by

$$K^i = \nabla_k t^{ki}. \quad (102.8)$$

103. Alternative form of the equations for continuous media. Equations (102.7) can be put into an interesting alternative form by writing out the acceleration components γ^i explicitly. These components are simply the absolute total derivatives with respect to time of the velocity vector **v**. Using (98.2), we have

$$\gamma^i = \frac{dv^i}{dt} + \Gamma_k{}^i{}_h v^k v^h = \frac{\partial v^i}{\partial t} + \Gamma_k{}^i{}_h v^h v^k + v^k \partial_k v^i,$$

or
$$\gamma^i = \frac{\partial v^i}{\partial t} + v^k \nabla_k v^i. \tag{103.1}$$

We deduce that

$$\rho\gamma^i = \frac{\partial(\rho v^i)}{\partial t} + \nabla_k(\rho v^k v^i) - v^i\left[\frac{\partial\rho}{\partial t} + \nabla_k(\rho v^k)\right],$$

and, since the bracketed term is zero by virtue of the equation of continuity,

$$\dot{\rho\gamma^i} = \frac{\partial(\rho v^i)}{\partial t} + \nabla_k(\rho v^k v^i).$$

We thus obtain a new formulation for the general dynamical equations for continuous media:

$$\frac{\partial(\rho v^i)}{\partial t} + \nabla_k(\rho v^k v^i + t^{ki}) = f^i. \tag{103.2}$$

It was this formulation which suggested some of the postulates of general relativity to Einstein.

The three equations (103.2) together with the equation of continuity determine the behaviour of continuous media under the action of surface and other forces. The mechanical problem is, of course, only defined when the t^{ik} and the f^i are known. The body forces are determined by external circumstances such as the presence of a gravitational field, whilst the stresses depend upon the internal deformations of the medium and on the flux of matter. In the case of a fluid, for example,

$$t_{ik} = p g_{ik}$$

where p is itself the function of ρ and temperature given by the equation of state of the fluid.

The above equations can be used to derive an equation representing energy transfer which expresses the fact that the energy contained in an element of volume varies with the flux of energy passing through the surface of the element.

CHAPTER VII

Special Relativity and Maxwell's Equations

I. PHYSICAL PRINCIPLES

104. The Michelson-Morley experiment. A systematic treatment of special relativity theory would take us beyond the scope of this book. In this chapter we merely propose to indicate the bases of the theory and to show the role played by tensor methods, particularly in the theory of electromagnetism.

Historically, the theory of special relativity was a consequence of the negative result of the Michelson-Morley experiment. Various experimental facts had pointed to the existence of an ether at absolute rest, which did not affect the motion of ordinary matter, and which was the medium in which electromagnetic waves were propagated. Among these facts was Fizeau's experiment on the velocity of propagation of light in a material medium and the phenomenon of aberration of light waves.

This picture of a stationary ether seemed to make it inevitable that the value of the speed of light measured by an observer in motion with respect to the ether would depend upon the magnitude and direction of his velocity. If c is the velocity of light and v is the collinear velocity of the observer with respect to the ether then, according to classical kinematics, the velocity measured by the observer would be $(c - v)$ or $(c + v)$ according to whether he moves in the same or the opposite direction as the light. An observer ignorant of his velocity relative to the ether would be able to determine this experimentally by emitting light signals in all directions and measuring the time taken by these signals to reach the surface of a sphere centred at the light source. Any motion with respect to the ether would create an 'ether wind' which would affect the signals in such a way that the first one to reach the sphere would have travelled along the direction opposite to the motion and the last one would have travelled in the same direction as the motion.

This was the principle of the celebrated Michelson-Morley experiment which attempted to find the motion of the earth with respect to the ether. The velocity of the earth in its orbit is about 30 km/sec. with respect to the Copernican reference frame which has its origin at the centre of gravity of the solar system. After six months this velocity is reversed. It might be that, at a certain instant, the unknown motion of the Copernican axes with respect to the ether exactly cancels the relative motion of the earth but this could hardly continue throughout the year.

Using a well-known interferometer whose description we shall omit, Michelson and Morley could have detected an ether drift of only 1·5 km/sec. In fact they observed no fringe shift within the limits of experimental accuracy and the same experiment done at different periods of the year always gave the same negative result. More recent experiments have confirmed the original result.†

Experiment thus shows that the speed of light is independent of the motion of the observer.

Various artificial hypotheses (the ether drag theory, Ritz' theory of the dependence of the speed of light on the motion of the source) were put forward to explain this negative result. These were subsequently vitiated by further experimental facts. However, it was the hypothesis of the contraction of bodies in motion in the quantitative form put forward by Fitzgerald which led physicists, notably Lorentz and Einstein, to formulate the special theory of relativity.

105. Constancy of the speed of light. Lorentz and Einstein took the results of the Michelson-Morley experiment as their starting point. Since a Galilean frame is a system of axes moving uniformly along a straight line with respect to the Copernican frame the result of this experiment can be stated as follows: the speed of light is the same with respect to all Galilean frames which coincide for short periods with the system of axes fixed on the earth as it describes its orbit.

We are thus led to postulate the *principle of the constancy of the speed of light*:

† These were by Kennedy (1926), Piccard and Stahel (1926–8), Joos (1930). Indirect confirmation was given by the observations of Ives (1938) on the Doppler effect.

The speed of light in vacuo measured with respect to any Galilean frame is the same in all directions; this speed, of the order of 300,000 km/sec, will be denoted by c.

There might be disadvantages in basing a principle of such generality on the result of a single type of experiment without the confirmation of other types of experiment. However, the Michelson-Morley experiment merely served to draw the attention of physicists to a mathematical fact which had, until then, remained obscure. It had already been pointed out by Poincaré that the equations of Newtonian dynamics and Maxwell's electromagnetic equations are not invariant under the same group of transformations. An intrinsic conflict therefore exists between Newtonian dynamics and electromagnetic theory.

In order to resolve this conflict Einstein proposed to admit the principle of the constancy of the speed of electromagnetic waves, thus retaining the Maxwell theory, and to modify Newtonian dynamics accordingly.

106. Newton's and Einstein's principles of relativity. In order to improve our understanding of the above-mentioned conflict let us return to Newtonian mechanics. Consider any two Galilean frames whose relative velocity is represented by v. We can arrange, by a suitable choice of coordinates, that the axis $O'x'$ of the second frame slides along the axis Ox of the first, the axes $O'y'$ and $O'z'$ remaining parallel respectively to the axes Oy and Oz of the first frame. Using classical kinematics, we have that, if $(xyzt)$ and $(x'y'z't')$ denote the coordinates and time of the same event with respect to the two Galilean frames, then

$$x' = x - vt, \qquad y' = y, \qquad z' = z, \qquad t' = t, \qquad (106.1)$$

where we have chosen the same origin for t and t'. Combining (106.1) with transformations which consist of pure spatial translations and rotations combined with changes of the time origin we obtain the most general transformation in classical mechanics, relating the coordinates and time of an event in a Galilean frame to the coordinates and time of the same event in any other Galilean frame. These transformations form a group which is known as the *classical Galilean group*.

In Newtonian dynamics the various material points M_1, M_2 are subject to forces which are derivable from a potential function which depends only on the relative distances r_{12}, r_{13} etc. of the points at the same instant of time. If γ_1 denotes the acceleration of M_1 with respect to a Galilean frame the corresponding equation of motion of M_1 may be written

$$m_1 \gamma_1 = \text{grad}_{(M_1)} \Phi, \text{etc.}; \qquad \Phi = \Phi(r_{12}, r_{13}, \text{etc.}).$$

Changing to another Galilean frame by means of a transformation in the classical Galilean group we see immediately that

$$\gamma_1 = \gamma_1', \text{etc.}, \qquad r_{12} = r_{12}', \text{etc.},$$

and in the new frame the equations of motion become

$$m_1 \gamma_1' = \text{grad}_{(M_1)} \Phi', \text{etc.}; \qquad \Phi' = \Phi(r_{12}', r_{13}', \text{etc.}).$$

The Newtonian equations of motion are thus invariant with respect to the classical Galilean group. We conclude that, *according to Newtonian mechanics, no purely mechanical experiment carried out within a Galilean system can determine the motion of this Galilean system with respect to any other Galilean system.*

This constitutes *Newton's principle of relativity*. This principle does not include electromagnetic phenomena since Maxwell's equations are not invariant with respect to the classical Galilean group.

The Michelson-Morley experiment and the principle of the constancy of the speed of light led to the formulation of *Einstein's principle of relativity:*

No physical experiment – mechanical or electromagnetic – carried out within a Galilean system can be used to demonstrate the motion of this Galilean system with respect to any other Galilean system.

The idea of an ether at absolute rest thus loses all significance and is eliminated. All the equations of physics, referred to Galilean frames, must be invariant with respect to certain transformations which, in particular, leave Maxwell's equations invariant. These transformations form a group, known as the *Lorentz group*, which we now propose to study.

II. THE LORENTZ GROUP AND
MINKOWSKI SPACE-TIME

107. Space-time. When we were studying dynamical systems with time-dependent constraints in the previous Chapter it was found convenient to use the configuration space-time V_{n+1}. Similarly, in order to represent mechanical and electromagnetic phenomena taking place in the world, it is natural to introduce a continuum V_4 having four dimensions, three spatial and one temporal, each point of which corresponds to an event. This continuum is called *space-time*.

The space-time continuum can be referred to arbitrary systems of curvilinear coordinates. If (y^1, y^2, y^3, y^4) denotes one such coordinate system it is possible to make coordinate transformations of the type

$$y^\alpha = f^\alpha(y^{1'}, y^{2'}, y^{3'}, y^{4'}) \quad (\alpha = 1, 2, 3, 4), \qquad (107.1)$$

where the f^α are restricted only to the extent of the usual conditions of continuity, of single-valuedness and of differentiability.

In special relativity theory we assume that it is possible to refer at least part of this continuum to a Galilean coordinate system defined in the following way: let (x, y, z) be the coordinates of a point in space with respect to a rectangular Galilean system of axes. The principle of the constancy of the speed of light provides us with a natural time scale which allows the definition of a time t associated with an event occurring at the point (x, y, z). The coordinates (x, y, z, t) are said to be the Galilean coordinates of V_4. We propose to find the transformation formulae

$$\left. \begin{aligned} x &= x(x'y'z't') \\ y &= y(x'y'z't') \\ z &= z(x'y'z't') \\ t &= t(x'y'z't') \end{aligned} \right\} \qquad \left. \begin{aligned} x' &= x'(xyzt) \\ y' &= y'(xyzt) \\ z' &= z'(xyzt) \\ t' &= t'(xyzt) \end{aligned} \right\} \qquad (107.2)$$

which enable us to pass from one system of Galilean coordinates (x, y, z, t) to another (x', y', z', t').

108. The Lorentz group. We now consider two Galilean coordinate systems and express the infinitesimal displacement of an electromagnetic wave with respect to each of them.

In terms of the first system of rectangular coordinates, the spatial line element is given by

$$d\sigma^2 = dx^2 + dy^2 + dz^2.$$

Consider, in this system, two neighbouring events (x, y, z, t) and $(x+dx, y+dy, z+dz, t+dt)$. If these two events define the infinitesimal displacement of an electromagnetic wave we have, according to the principle of the constancy of the speed of light,

$$\frac{d\sigma^2}{dt^2} = c^2.$$

Hence, for any infinitesimal displacement of an electromagnetic wave, the expression

$$ds^2 = c^2 dt^2 - d\sigma^2 = c^2 dt^2 - dx^2 - dy^2 - dz^2 \qquad (108.1)$$

is zero. The same expression, evaluated in the second coordinate system,

$$ds'^2 = c^2 dt'^2 - d\sigma'^2 = c^2 dt'^2 - dx'^2 - dy'^2 - dz'^2$$

is also zero. Consider now an *arbitrary* infinitesimal displacement at a point M in the two systems. A reasonable generalization of the relation $ds'^2 = 0 = ds^2$ is given by

$$ds'^2 = f(M)ds^2,$$

where $f(M)$ is a function of the coordinates of M and may also depend on the relative velocity $\mathbf{v}$ of the two coordinate systems. More explicitly

$$c^2 dt'^2 - dx'^2 - dy'^2 - dz'^2 = f(x, y, z, t)(c^2 dt^2 - dx^2 - dy^2 - dz^2).$$
$$(108.3)$$

When the relative velocity of the two Galilean systems is zero the function $f(M)$ must reduce to unity. It can be shown that the only physically significant solutions of (108.3) are those in which $f(M)$ is always unity. Equation (108.3) therefore becomes

$$c^2 dt'^2 - dx'^2 - dy'^2 - dz'^2 = c^2 dt^2 - dx^2 - dy^2 - dz^2. \quad (108.4)$$

We are thus led to consider all coordinate transformations (107.2) which satisfy the condition (108.4) and therefore leave invariant the differential quadratic form defined by (108.1). We also require them to reduce to a transformation consisting of a pure spatial displacement and a change of time origin when $v = 0$. It is immediately obvious that these transformations form a group: if any two transformations (107.2) leave the quadratic form (108.1) invariant, so does their product; the inverse of such a transformation also leaves (108.1) invariant. This group of transformations is known as the *Lorentz group*.

The determination of the transformations of the Lorentz group is a simple algebraic problem, the solution of which is outlined below. A geometrical interpretation of the Lorentz group can also easily be obtained. Put

$$u = ict, \qquad u' = ict', \qquad (i^2 = -1), \qquad (108.5)$$

then (108.4) takes the form

$$dx'^2 + dy'^2 + dz'^2 + du'^2 = dx^2 + dy^2 + dz^2 + du^2. \qquad (108.6)$$

The Lorentz group thus corresponds to the group of translations and rotations in a proper Euclidean space of four dimensions. This can be obtained by combining the translations

$$x' = x + a, \qquad y' = y + b, \qquad z' = z + c, \qquad u' = u + d,$$

with the linear orthogonal transformations (rotations) of the four coordinates (x, y, z, u).

According to its definition the Lorentz group contains transformations which consist of purely spatial displacements (which conserve the form $dx^2 + dy^2 + dz^2$) and changes of the time origin. We now seek a particular transformation based on the trivial solutions already obtained which generates the most general transformation of the Lorentz group.

To this end we note that by adding a constant to each of the four variables, which is equivalent to a translation of the axes $Oxyz$ and a change in the origin of the time t, the resulting functions (107.2) which are necessarily linear, can also be made homogeneous. This implies that the origin of the first frame at time $t = 0$ coincides with that of the second frame at $t' = 0$. In a rectilinear translation of the second frame with respect to the first, a straight line fixed in the second frame and parallel to the relative velocity slides along a straight line fixed in the first frame. By a suitable choice of the spatial coordinates we can arrange that the $O'x'$ axis slides along Ox in the course of the motion, the directions of these two axes then being that of the relative velocity. Rotating $O'x'y'z'$ about $O'x'$ we can make $O'y'$ and $O'z'$ parallel and in the same sense as Oy and Oz respectively. This means that $Oxyz$ at time $t = 0$ coincides with $O'x'y'z'$ at time $t' = 0$.

Returning to the four-dimensional Euclidean space referred to the orthonormal frames with coordinates (x, y, z, u) and (x', y', z', u'), we are thus led to study rotations which simultaneously leave invariant the plane defined by $y = 0$ or $y' = 0$ and the one defined by $z = 0$ or $z' = 0$; that is to say they leave invariant the orthogonal coordinate axes corresponding to the variables z and y. We conclude, as in the three-dimensional case, that

$$y' = y, \qquad z' = z. \tag{108.7}$$

The identity (108.6) now reduces to

$$dx'^2 + du'^2 = dx^2 + du^2, \tag{108.8}$$

and the transformation only concerns the two variables x and u. Since the origins of the coordinates (x, u) and (x', u') coincide, the identity (108.8) defines ordinary rotations in a plane

$$\left.\begin{array}{l} x' = x\cos\phi + u\sin\phi, \\ u' = -x\sin\phi + u\cos\phi, \end{array}\right\} \tag{108.9}$$

where ϕ denotes an arbitrary angular parameter. Since u and u' are pure imaginary quantities, $\sin\phi$ is pure imaginary and $\cos\phi$ is real. As $\tan\phi$ is pure imaginary and dimensionless we write

$$\tan\phi = \frac{iv}{c},$$

and substitute for ϕ the real parameter v having the dimensions of velocity. It follows that

$$\cos\phi = \frac{1}{\sqrt{(1-v^2/c^2)}}, \qquad \sin\phi = \frac{i(v/c)}{\sqrt{(1-v^2/c^2)}}. \quad (108.10)$$

Substituting these expressions into (108.9) and reintroducing the variables t and t', we obtain

$$x' = \frac{x-vt}{\sqrt{(1-v^2/c^2)}}, \qquad t' = \frac{t-vx/c^2}{\sqrt{(1-v^2/c^2)}}.$$

We have thus established that the most general transformation of the Lorentz group is obtained by combining the trivial transformations (pure spatial rotations and translations) with the special transformation

$$x' = \frac{x-vt}{\sqrt{(1-\beta^2)}}, \quad y' = y, \quad z' = z, \quad t' = \frac{t-\beta(x/c)}{\sqrt{(1-\beta^2)}}, \quad \left(\beta = \frac{v}{c}\right).$$
$$(108.11)$$

This is known as the special Lorentz transformation.

It is obvious from (108.11) that v which, for physical reasons is *necessarily smaller than c* in absolute value, is such that every point fixed in the second Galilean frame (the origin, for example) is moving relative to the first frame with uniform velocity parallel to Ox. We note that if $\beta = v/c$ becomes negligibly small (108.11) reduces to (106.1) which was used to define the classical Galilean group. When $v = 0$ the transformation (108.11) naturally reduces to the identity transformation. The inverse formulae to those in (108.11) are given by

$$x = \frac{x'+vt'}{\sqrt{(1-\beta^2)}}, \qquad y = y', \qquad z = z', \qquad t = \frac{t'+\beta(x'/c)}{\sqrt{(1-\beta^2)}},$$
$$(108.12)$$

which only differ by the interchange of variables and the replacement of v by $-v$.

It would take rather long to verify directly that Maxwell's equations, in their usual form, are invariant with respect to transformations of the Lorentz group. This result is, however, a trivial consequence of the simple tensor form into which we shall now put these equations.

109. Lorentz transformation in vector form. It is convenient to rewrite the special Lorentz transformation in three-dimensional vector notation.

We denote the velocity of the second frame with respect to the first by **v** and the unit vector along Ox (and in the direction of **v**) by **i**, hence

$$\mathbf{v} = v\mathbf{i}.$$

Let **r** represent the position vector defining the point (x, y, z) with respect to the origin O of the first frame and **r**′ represent the analogous quantity in the second frame. The vectors **r** and **r**′ may be written as

$$\left.\begin{array}{l} \mathbf{r} = x\mathbf{i} + \mathbf{r}_1, \\ \mathbf{r}' = x'\mathbf{i} + \mathbf{r}_1', \end{array}\right\} \tag{109.1}$$

where $\mathbf{r}_1$ and $\mathbf{r}_1'$ represent the components of **r** and **r**′ perpendicular to **i**. Using the Lorentz transformation relations (108.12) we have

$$x = \frac{x' + vt}{\sqrt{(1 - \beta^2)}}, \qquad \mathbf{r}_1 = \mathbf{r}_1'.$$

We now wish to evaluate **r** in terms of **r**′, **v** and t'. Combining equations (109.1) we have

$$\mathbf{r} = \mathbf{r}' + (x - x')\mathbf{i}. \tag{109.2}$$

However

$$x - x' = \frac{x' + vt'}{\sqrt{(1 - \beta^2)}} - x' = \left(\frac{1}{\sqrt{(1 - \beta^2)}} - 1\right)x' + \frac{vt'}{\sqrt{(1 - \beta^2)}},$$

and

$$x'\mathbf{i} = (\mathbf{r}' \cdot \mathbf{i})\mathbf{i} = (\mathbf{r}' \cdot \mathbf{v})\frac{\mathbf{v}}{v^2}.$$

Substituting these results into (109.2) we get the transformation formula for position vectors:

$$\mathbf{r} = \mathbf{r}' + \left(\frac{1}{\sqrt{(1 - \beta^2)}} - 1\right)\mathbf{r}' \cdot \mathbf{v}\frac{\mathbf{v}}{v^2} + \frac{vt'}{\sqrt{(1 - \beta^2)}}. \tag{109.3}$$

It is apparent that the three vectors **r**, **r**′ and **v** are always coplanar.

Using (108.12) the time transformation formula can obviously be written

$$t = \frac{t' + (\mathbf{r}' \cdot \mathbf{v})/c^2}{\sqrt{(1 - \beta^2)}}.$$ (109.4)

The transformation equations in their vector form (109.3) and (109.4) are useful in a great number of problems in special relativity. We shall now use them to establish Einstein's formula for the composition of velocities.

110. Relativistic composition of velocities. The composition of velocities in classical kinematics is a simple matter: if $\mathbf{v}$ denotes the velocity of a Galilean system S' with respect to S and if a material point P moves with velocity $\mathbf{u}'$ relative to S' then its velocity relative to S is given by

$$\mathbf{u} = \mathbf{v} + \mathbf{u}'.$$ (110.1)

In special relativity the relation between $\mathbf{u}$, $\mathbf{v}$ and $\mathbf{u}'$ is much more complicated. Using the notation of §109 the velocities of P with respect to the two Galilean systems are given by

$$\mathbf{u} = \frac{d\mathbf{r}}{dt}, \qquad \mathbf{u}' = \frac{d\mathbf{r}'}{dt'}.$$

Differentiating (109.3) with respect to t we obtain

$$\frac{d\mathbf{r}}{dt} = \left[\frac{d\mathbf{r}'}{dt'} + \left(\frac{1}{\sqrt{(1 - \beta^2)}} - 1\right)\left(\frac{d\mathbf{r}'}{dt'} \cdot \mathbf{v}\right)\frac{\mathbf{v}}{v^2} + \frac{\mathbf{v}}{\sqrt{(1 - \beta^2)}}\right]\frac{dt'}{dt}, \quad (110.2)$$

and differentiating (109.4) with respect to t' gives

$$\frac{dt}{dt'} = \frac{1 + (\mathbf{u}' \cdot \mathbf{v})/c^2}{\sqrt{(1 - \beta^2)}}.$$

Substituting this expression for dt/dt' in (110.2) we find

$$\mathbf{u} = \frac{1}{1 + (\mathbf{u}' \cdot \mathbf{v}/c^2)}\left(\sqrt{(1 - \beta^2)}\,\mathbf{u}' + [1 - \sqrt{(1 - \beta^2)}]\frac{\mathbf{u}' \cdot \mathbf{v}}{v^2}\mathbf{v} + \mathbf{v}\right),$$

which shows that the three vectors $\mathbf{u}$, $\mathbf{u}'$ and $\mathbf{v}$ are coplanar. This expression is easily transformed into the formula for the composition of velocities, giving

$$\mathbf{u} = \frac{1}{1+(\mathbf{u}'\cdot\mathbf{v}/c^2)}\left[\left(1+\frac{\mathbf{u}'\cdot\mathbf{v}}{v^2}\right)\mathbf{v}+\sqrt{(1-\beta^2)}\left(\mathbf{u}'-\frac{\mathbf{u}'\cdot\mathbf{v}}{v^2}\mathbf{v}\right)\right], \quad (110.3)$$

in which the numerator is the sum of two vectors, one parallel and one perpendicular to $\mathbf{v}$. If $\beta = v/c$ is negligible compared to unity, equation (110.3) has the limiting form of the classical relation (110.1). An immediate consequence of (110.3) is the expression for the square of $\mathbf{u}$

$$u^2 = \frac{1}{[1+(\mathbf{u}'\cdot\mathbf{v}/c^2)]^2}\left[u'^2+v^2+2\mathbf{u}'\cdot\mathbf{v}+\frac{(\mathbf{u}'\cdot\mathbf{v})^2}{c^2}-\frac{u'^2v^2}{c^2}\right], \quad (110.4)$$

which is symmetrical in $\mathbf{u}'$ and $\mathbf{v}$.

When the vectors $\mathbf{u}'$ and $\mathbf{v}$ are collinear $\mathbf{u}$ is also in the same direction and the second term in the numerator of (110.3) is zero. We then have the Einstein relation between the algebraic values of the three velocities

$$u = \frac{u'+v}{1+(u'v/c^2)}. \quad (110.5)$$

It is easily seen from (110.4) or (110.5) that it is impossible to exceed the speed of light by combining two velocities less than or equal to c. In particular putting $u' = c$ in (110.4) gives

$$u^2 = \frac{c^2}{[1+(\mathbf{u}'\cdot\mathbf{v}/c^2)]^2}\left(1+\frac{\mathbf{u}'\cdot\mathbf{v}}{c^2}\right)^2 = c^2.$$

Thus, when $u' = c$, it always follows that $u = c$ whatever the value of $\mathbf{v}$. This result obviously agrees with the principle of the constancy of the speed of light.

111. Minkowski space-time. Equation (108.4) is equivalent to saying that two neighbouring events (x,y,z,t) and $(x+dx,y+dy,z+dz,t+dt)$ define an expression

$$ds^2 = c^2\,dt^2-dx^2-dy^2-dz^2, \quad (111.1)$$

which has the same value in all Galilean coordinate systems. We note that, if the two events correspond to a displacement made with a speed less than c, then ds^2 is positive.

We are thus led, in special relativity, to associate the continuum V_4 with the metric defined by the quadratic form (111.1). Since this metric has constant coefficients with respect to the Galilean variables and a signature $(+ - - -)$, it therefore defines V_4 as an *improper Euclidean space*. This space is known as Minkowski space-time and ds is said to be the infinitesimal element of a finite *interval* between two events.

The Galilean coordinates constitute an orthogonal rectilinear coordinate system for this space. We obtain an orthonormal frame in V_4 if we substitute the variables

$$x^1 = x, \qquad x^2 = y, \qquad x^3 = z, \qquad x^4 = ct, \qquad (111.2)$$

for (x, y, z, t). The metric (111.1) then takes the form

$$ds^2 = (dx^4)^2 - (dx^1)^2 - (dx^2)^2 - (dx^3)^2.$$

The variables x^α are called *reduced Galilean coordinates*. We shall make frequent use of them in what follows. It is convenient to introduce the notation

$$ds^2 = \eta_{\alpha\beta} dx^\alpha dx^\beta \quad (\alpha, \beta, \text{etc.} = 1, 2, 3, 4) \qquad (111.3)$$

where

$$\eta_{\alpha\beta} = 0 \quad (\alpha \neq \beta), \qquad \eta_{44} = -\eta_{11} = -\eta_{22} = -\eta_{33} = 1.$$

Naturally, Minkowski space-time can be referred to any system of curvilinear coordinates. Its metric is then written as

$$ds^2 = g_{\alpha\beta} dy^\alpha dy^\beta$$

where the $g_{\alpha\beta}$ are functions of the curvilinear coordinates y^α.

III. DYNAMICS IN SPECIAL RELATIVITY

112. The unit velocity vector and the principle of inertia. Consider a moving mass point P in Minkowski space-time whose velocity is, of course, less than c. The motion of P can be defined by giving the coordinates y^α as functions of a single parameter. We choose the interval s measured along the trajectory of P in V_4. The vector

$$u^\alpha = \frac{dy^\alpha}{ds} \quad (\alpha = 1, 2, 3, 4) \tag{112.1}$$

which is obviously of unit length, is called the *unit velocity vector*.

Let us express the components of this vector in terms of reduced Galilean coordinates. The components of an ordinary spatial velocity vector are

$$v^i = \frac{dx^i}{dt} \quad (i, \text{ etc.} = 1, 2, 3). \tag{112.2}$$

We conclude that

$$\left.\begin{aligned}
u^i &= \frac{dx^i}{ds} = v^i \frac{dt}{ds}, \\
u^4 &= \frac{dx^4}{ds} = c \frac{dt}{ds}.
\end{aligned}\right\}$$

However, if $v = \beta c$ denotes the magnitude of the velocity (112.2), we can rearrange (111.1) to give

$$\frac{dt}{ds} = \frac{1}{\sqrt{(c^2 - v^2)}} = \frac{1}{c\sqrt{(1 - \beta^2)}}. \tag{112.3}$$

It follows that

$$u^i = \frac{v^i}{c\sqrt{(1 - \beta^2)}}, \qquad u^4 = \frac{1}{\sqrt{(1 - \beta^2)}}. \tag{112.4}$$

Let the motion of P in Galilean coordinates be uniform and rectilinear. The components v^i, and consequently the u^α, are then constant. We therefore have

$$\frac{du^i}{ds} = 0, \qquad \frac{du^4}{ds} = 0. \tag{112.5}$$

The world line of P is therefore a geodesic of (111.1) with positive ds^2. On the other hand the differential system describing the geodesics of (111.1) may be written in Galilean variables as

$$\frac{d^2 x}{dt^2} = 0, \qquad \frac{d^2 y}{dt^2} = 0, \qquad \frac{d^2 z}{dt^2} = 0,$$

and each geodesic with positive ds^2 corresponds to a uniform rectilinear motion with velocity less than c.

In Newtonian dynamics, the principle of inertia asserts that an isolated mass point is always in uniform rectilinear motion with respect to Galilean axes. In special relativity we preserve this principle and reformulate it in the following invariant manner:

Principle of inertia: An isolated mass point follows a path which is a geodesic with positive ds^2 in Minkowski space-time.

Geodesics with $ds^2 = 0$ correspond to straight lines traversed with velocity c, that is to the paths followed by light rays. We thus see that a simple relation exists between special relativity and the geometry of Minkowski space-time.

113. The equations of motion. Consider the motion of a mass point P under the action of a known force $\mathbf{f}$ which varies with the position and velocity of P. According to Newtonian dynamics we have

$$m\frac{d\mathbf{v}}{dt} = \mathbf{f}, \tag{113.1}$$

where m denotes the mass of the particle and $\mathbf{v}$ is its velocity vector in space. Using (113.1) we obtain the energy equation

$$\frac{d}{dt}(\tfrac{1}{2}mv^2) = \mathbf{f}\cdot\mathbf{v}. \tag{113.2}$$

The Newtonian equations (113.1) and (113.2) are not invariant under the Lorentz group which transforms from one Galilean coordinate system to another in relativity theory. They must therefore be modified in order to comply with the demands of special relativity. In this process we shall be guided by the requirement that the new equations should reduce to the old ones for small velocities.

Equations (112.5) represent the motion of an isolated mass point in relativistic form. They suggest that the relativistic equations of motion should be formulated in terms of du^α/ds rather than the classical acceleration vector. In arbitrary curvilinear coordinates the relativistic acceleration is given by the expression

$$J^\alpha = \frac{\nabla u^\alpha}{ds} \quad (\alpha = 1, 2, 3, 4).$$

Let us characterize the inertia of a point P by a parameter m_0, having the dimensions of mass, which we shall call the *rest-mass* of P. We postulate the following form of the equations of motion

$$m_0 c^2 \frac{\nabla u^\alpha}{ds} = \Phi^\alpha, \tag{113.3}$$

where Φ^α is the relativistic generalization of the Newtonian force vector. The coefficient c^2 has been introduced for reasons which will appear shortly. The equations (113.3), which are the equations of motion in special relativity, were first given by Minkowski.

Since u^α is a unit vector its absolute differential ∇u^α is orthogonal to it. Then according to (113.3)

$$\Phi^\alpha u_\alpha = 0. \tag{113.4}$$

Thus the vector Φ^α is always orthogonal to the unit velocity vector.

We now use Galilean coordinates and interpret equations (113.3) by comparison with the Newtonian equations (113.1) and (113.2). In such a coordinate system equation (113.3) can be written

$$m_0 c^2 \frac{du^\alpha}{ds} = \Phi^\alpha,$$

and reintroducing the variable t, related to s by (112.3), we have

$$m_0 c \frac{du^\alpha}{dt} = \Phi^\alpha \sqrt{(1-\beta^2)}.$$

Using (112.4) to evaluate the components u^α as functions of the components of the spatial velocity $\mathbf{v}$ we find

$$m_0 \frac{d}{dt}\left(\frac{v^i}{\sqrt{(1-\beta^2)}}\right) = \Phi^i \sqrt{(1-\beta^2)}, \tag{113.5}$$

$$m_0 \frac{d}{dt}\left(\frac{c}{\sqrt{(1-\beta^2)}}\right) = \Phi^4 \sqrt{(1-\beta^2)}, \tag{113.6}$$

which bear a simple relationship to the Newtonian equations. Denoting the spatial vector whose components are

$$f^i = \Phi^i \sqrt{(1-\beta^2)}, \tag{113.7}$$

by $\mathbf{f}$, we may rewrite equations (113.5) in ordinary vector notation as

$$m_0 \frac{d}{dt}\left(\frac{\mathbf{v}}{\sqrt{(1-\beta^2)}}\right) = \mathbf{f}, \tag{113.8}$$

which recalls the fundamental equation of Newtonian dynamics (113.1) and reduces to it when β becomes negligible compared to unity.

It is a little more complicated to interpret the right-hand side of (113.6). The orthogonality condition (113.4) may be written in reduced Galilean coordinates as

$$\Phi^4 u^4 = \sum_i \Phi^i u^i,$$

or using (112.4) and (113.7),

$$\frac{\Phi^4}{\sqrt{(1-\beta^2)}} = \sum_i \frac{f^i v^i}{c(1-\beta^2)} = \frac{\mathbf{f}\cdot\mathbf{v}}{c(1-\beta^2)},$$

from which

$$\Phi^4 \sqrt{(1-\beta^2)} = \frac{\mathbf{f}\cdot\mathbf{v}}{c}.$$

Then (113.6) can be written

$$\frac{d}{dt}\left(\frac{m_0 c^2}{\sqrt{(1-\beta^2)}}\right) = \mathbf{f}\cdot\mathbf{v}. \tag{113.9}$$

The right-hand side of this equation is identical with that of the classical mechanical relation (113.2); so the time variation of the quantity

$$E = \frac{m_0 c^2}{\sqrt{(1-\beta^2)}} \tag{113.10}$$

is equal to the work done by the force $\mathbf{f}$. We are therefore led to suppose that (113.10) defines the total energy of the mass point. It is important to note that this energy does not reduce to zero as v tends to zero. We write

$$E = E_0 + T,$$

where

$$E_0 = m_0 c^2, \qquad T = m_0 c^2\left[\frac{1}{\sqrt{(1-\beta^2)}} - 1\right]. \tag{113.11}$$

We shall call E_0 the *rest energy* of the mass point and T the *relativistic kinetic energy*. For small values of v

$$T = \tfrac{1}{2}m_0 c^2 \beta^2 = \tfrac{1}{2}m_0 v^2.$$

Hence, for small velocities, E differs only by the constant E_0 from the ordinary kinetic energy of classical mechanics.

114. Momentum-energy vector and relativistic mass. The vector

$$p^\alpha = m_0 c u^\alpha, \qquad (114.1)$$

which is collinear with the unit velocity vector, is called the *momentum-energy vector*. According to (112.4) the components of this vector are given by

$$p^i = m_0 \frac{v^i}{\sqrt{(1-\beta^2)}}, \qquad p^4 = m_0 \frac{c}{\sqrt{(1-\beta^2)}} = \frac{E}{c}, \qquad (114.2)$$

in reduced Galilean coordinates.

Introducing the momentum vector **p** whose components are p^i, the fundamental equation (113.8) may be written in the form

$$\frac{d\mathbf{p}}{dt} = \mathbf{f}. \qquad (114.3)$$

This can be obtained from the Newtonian equation of motion (113.1) by replacing $m\mathbf{v}$ by **p**. Thus the point P can be considered to have the variable mass

$$m = \frac{m_0}{\sqrt{(1-\beta^2)}}. \qquad (114.4)$$

This quantity m is known as the *relativistic mass* of P. For small velocities m reduces to the rest mass m_0 and, as v approaches the velocity of light, m tends to infinity.

According to (113.10) the *relativistic mass* is equal to the *total energy* E divided by c^2 while the *rest mass* is the *rest energy* divided by c^2. It thus appears that there is a direct correlation between mass and energy in relativity theory which has no analogue in classical physics. We now propose to investigate this relationship.

115. The inertia of energy. Consider a complex material particle referred to a Galilean frame S_0 which is chosen in such a way that the total momentum of the particle is zero. Suppose also that the particle has the total energy E_0 relative to S_0; this is its *rest energy*.† The momentum-energy vector of the particle then has the components

$$p_0^i = 0, \qquad p_0^4 = \frac{E_0}{c^2},$$

with respect to S_0.

Now consider another Galilean frame S moving with velocity **v** relative to S_0. Carrying out a Lorentz transformation, we see that the components of the momentum-energy vector with respect to S are given by

$$p^i = \frac{E_0}{c^2} \frac{v^i}{\sqrt{(1-\beta^2)}}, \qquad p^4 = \frac{E_0}{c^2} \frac{c}{\sqrt{(1-\beta^2)}}. \tag{115.1}$$

It is clear from these relations that all the rest energy E_0 makes a contribution to the spatial momentum vector as if it were a mass E_0/c^2. We are thus led to formulate the following principle:

A system having any form of rest energy E_0 has a corresponding inertial mass

$$m_0 = \frac{E_0}{c^2}. \tag{115.2}$$

This statement, which is known as the *principle of inertia of energy*, is due to Einstein. The recognition of the equivalence of mass and energy is, without doubt, the most fertile contribution that the special theory of relativity has made to physics.

As an example of this principle let us suppose that the particle, referred to S_0, is not subject to any forces and emits an energy ΔE_0 in the form of electromagnetic radiation during a finite interval of time. This might well be in the form of a spherical wave and the total momentum with respect to S_0 of the energy radiated would then be zero. It follows that the particle, initially at rest with respect to S_0, will remain so after emitting the radiation.

† This includes the interaction energy of the components of the 'particle' as well as the energy which corresponds to their masses. (T.)

Now refer the description to the S frame. Since the particle and the radiation constitute an isolated system, there will be conservation of the total momentum-energy vector. Hence, the radiation has the momentum

$$\frac{\Delta E_0}{c^2} \frac{\mathbf{v}}{\sqrt{(1-\beta^2)}}$$

and the energy

$$\frac{\Delta E_0}{\sqrt{(1-\beta^2)}}$$

with respect to S. The particle has thus supplied this momentum and energy without changing its velocity. This is only possible if its rest-mass is changed. If m_0 denotes the rest-mass of the particle before the emission and m_0' that after emission then we have the conservation relations

$$\frac{m_0 \mathbf{v}}{\sqrt{(1-\beta^2)}} = \frac{m_0' \mathbf{v}}{\sqrt{(1-\beta^2)}} + \frac{\Delta E_0}{c^2} \frac{\mathbf{v}}{\sqrt{(1-\beta^2)}},$$

$$\frac{m_0 c^2}{\sqrt{(1-\beta^2)}} = \frac{m_0' c^2}{\sqrt{(1-\beta^2)}} + \frac{\Delta E_0}{\sqrt{(1-\beta^2)}},$$

which are both equivalent to

$$m_0' = m_0 - \frac{\Delta E_0}{c^2}.$$

The mass of the particle has thus been decreased by an amount equal to the energy emitted divided by c^2.

In Newtonian mechanics there are separate conservation laws for the mass and energy of an isolated system. In relativistic mechanics, however, there is a single law of conservation for the total energy of an isolated system and the rest-masses of the constituents will alter every time kinetic energy is transformed into other forms of energy, and vice-versa. The rest-mass of a material particle remains constant as long as its rest-energy is not altered by such changes. But the rest-masses can change appreciably if the interaction energies are of the same order of magnitude as the rest energies. In this way it is possible to explain the mass defects observed in atomic nuclei and, in a general way, the mass losses which occur in most nuclear reactions.

The inertia of energy plays an important part in the understanding of nuclear phenomena, which are the object of so much research in contemporary physics. In these, energy is both materialized in the form of elementary particles and liberated by the disintegration of matter.

IV. RELATIVISTIC DYNAMICS OF CONTINUOUS MEDIA

116. Equations for a system at rest. In order to obtain the relativistic form of the equations of motion for a continuous medium, we introduce an orthogonal Galilean frame S_0 which is at rest with respect to some point P_0 of the medium. We then write down the equations at P_0 in terms of S_0.

At P_0 all the components v^i of the spatial velocity vector **v** are zero; on the other hand the derivatives of these components are generally different from zero. Denoting the unit velocity vector associated with v^i by u^α, it is clear that

$$u^i = 0, \qquad u^4 = 1 \tag{116.1}$$

at P_0. It follows, using (112.4), that the derivatives of the u^α are given by

$$\partial_\lambda u^i = \frac{1}{c} \partial_\lambda v^i, \qquad \partial_\lambda u^4 = 0, \tag{116.2}$$

since $\beta = 0$ at P_0 and

$$u^\alpha \partial_\lambda u_\alpha = 0,$$

because u^α is a unit vector.

The non-relativistic equations (99.2) and (103.2) take the form

$$\left. \begin{aligned} \frac{\partial \bar{p}^i}{\partial t} + \partial_k t^{ik} &= f^i, \\[2mm] \frac{\partial \rho}{\partial t} + \partial_k \bar{p}^k &= 0, \end{aligned} \right\} \tag{116.3}$$

in the system S_0 and at the point P_0, where $\bar{p}^i$ denotes the ordinary momentum vector per unit volume.

Let us consider equations (116.3) from the point of view of mass-energy equivalence. The ordinary momentum vector $\bar{p}^i = \rho v^i$ involves only the energy corresponding to the matter density ρ. However we know that in relativistic dynamics each form of energy must make its contribution to the momentum vector. Here we shall consider, in addition, only that energy which derives from the action of mechanical stresses, avoiding, in particular, the case where an electromagnetic interaction also occurs.

In order to evaluate the corresponding flux of energy, consider an element of area dS whose 'components' are denoted by the $d\sigma_k$ defined by (101.2). From (101.3) there is a corresponding surface force

$$T^l dS = t^{kl} d\sigma_k.$$

If the matter in the neighbourhood of this element of area moves with the velocity **v**, having contravariant components v^i, the corresponding work done is given by

$$\sum_l v^l t^{kl} d\sigma_k.$$

Introducing the covariant components of the velocity by means of the spatial metric $-(dx^1)^2 - (dx^2)^2 - (dx^3)^2$ this becomes

$$-v_l t^{kl} d\sigma_k.$$

In other words, a flux of energy traverses the element of area dS and the vector defining this total energy flux has the components

$$\rho c^2 v^i - v_l t^{il}, \qquad \bullet$$

where the first term corresponds to the density of matter ρ. The relativistic relation between energy and mass [see (115.1)] implies that the momentum vector per unit volume is the vector given above divided by c^2:

$$p^i = \rho v^i - \frac{1}{c^2} v_l t^{il}. \tag{116.4}$$

The components of this vector are zero at P_0, but this is not true of their derivatives which alone are involved in the equations of motion. In order to obtain the relativistic equations of motion valid at P_0 in terms of the frame S_0 we substitute (116.4) for the vector $\bar{\mathbf{p}}$ in the classical relations (116.3). We thus have

$$\left.\begin{aligned}
\frac{\partial}{\partial t}\left(\rho v^i - \frac{1}{c^2}v_l t^{il}\right) + \partial_k t^{ik} &= f^i, \\
\frac{\partial \rho}{\partial t} + \partial_k\left(\rho v^k - \frac{1}{c^2}v_l t^{kl}\right) &= 0,
\end{aligned}\right\}$$

and allowing for the fact that the v^i vanish at P_0,

$$\left.\begin{aligned}
\rho \frac{\partial v^i}{\partial t} - \frac{1}{c^2}\frac{\partial v_l}{\partial t}t^{il} + \partial_k t^{ik} &= f^i, \\
\frac{\partial \rho}{\partial t} + \rho \, \partial_k v^k - \frac{1}{c^2}(\partial_k v_l)\, t^{kl} &= 0.
\end{aligned}\right\}$$

Let us now replace the derivatives with respect to t by those with respect to $x^4 = ct$. We obtain, with the usual notation for partial derivatives,

$$\left.\begin{aligned}
\rho c \, \partial_4 v^i - \frac{1}{c}\partial_4 v_l t^{il} + \partial_k t^{ik} &= f^i, \\
c \, \partial_4 \rho + \rho \, \partial_k v^k - \frac{1}{c^2}\partial_k v_l t^{kl} &= 0,
\end{aligned}\right\} \tag{116.5}$$

which are the required equations of motion.

117. The equations of motion in tensor form. Starting from equations (116.5) which are valid for the point P_0 with respect to the frame S_0 we can determine the general tensor form of the equations of motion, thus translating the classical equations (99.2) and (103.2) into relativistic form.

The appearance of the tensor t^{ik} in the non-relativistic equations (103.2) suggests that a corresponding symmetrical second-order tensor must play an important role in the desired relativistic equations. We therefore introduce the tensor $T^{\lambda\mu}$ which has the components

$$T^{ik} = t^{ik}, \qquad T^{i4} = T^{4i} = T^{44} = 0,$$

at the point P_0 in the frame S_0, and which consequently satisfies the tensor equation

$$T^{\lambda\mu} u_\mu = 0. \qquad (117.1)$$

We also introduce the vector Φ^λ which has the components

$$\Phi^i = f^i, \qquad \Phi^4 = 0,$$

at the point P_0 in the frame S_0, and which is therefore orthogonal to the unit velocity vector:

$$\Phi^\lambda u_\lambda = 0. \qquad (117.2)$$

We now verify that equations (116.5) can be put into the following tensor form which is valid in an arbitrary curvilinear coordinate system:

$$\nabla_\mu (\rho c^2 u^\lambda u^\mu + T^{\lambda\mu}) = \Phi^\lambda. \qquad (117.3)$$

In fact, using the Galilean coordinate system S_0 at the point P_0, equations (117.3) reduce to

$$c^2 \rho \, \partial_4 u^i + \partial_k t^{ik} + \partial_4 T^{i4} = f^i \quad (\lambda = i), \qquad (117.4)$$

$$c^2 \partial_4 \rho + c^2 \rho \, \partial_k u^k + \partial_k T^{4k} + \partial_4 T^{44} = 0 \quad (\lambda = 4), \qquad (117.5)$$

when account is taken of (116.1) and (116.2). It is easy to evaluate the terms $\partial_\lambda T^{4\mu}$ appearing in (117.4) and (117.5). We have at P_0

$$\partial_\lambda T^{4\mu} = \partial_\lambda (u_4 T^{4\mu}).$$

Then using (117.1) we have

$$\partial_\lambda (u_4 T^{4\mu}) = -\partial_\lambda (u_l T^{l\mu}) = -\partial_\lambda u_l T^{l\mu},$$

so that $\qquad\qquad \partial_\lambda T^{4\mu} = -\partial_\lambda u_l T^{l\mu}.$

Substituting this expression for $\partial_\lambda T^{4\mu}$ into (117.4) and (117.5) we obtain

$$\left.\begin{array}{l} c^2 \rho \, \partial_4 u^i + \partial_k t^{ik} - \partial_4 u_l t^{li} = f^i, \\ c^2 \partial_4 \rho + c^2 \rho \, \partial_k u^k - \partial_k u_l t^{kl} = 0. \end{array}\right\} \qquad (117.6)$$

Then using equations (116.2) we recover (116.5). Equations (117.3) are the fundamental equations of relativistic dynamics for continuous media. $T^{\lambda\mu}$ is called the relativistic *stress tensor*.

118. The momentum-energy tensor. The form of the fundamental equations (117.3) leads us to introduce the symmetric tensor

$$P^{\lambda\mu} = \rho u^\lambda u^\mu + \frac{1}{c^2} T^{\lambda\mu}, \tag{118.1}$$

so that they may be rewritten as

$$\nabla_\mu P^{\lambda\mu} = \frac{1}{c^2} \Phi^\lambda. \tag{118.2}$$

$P^{\lambda\mu}$ is called the *momentum-energy tensor* of the continuous medium. We note that, because of (117.1),

$$P^{\lambda\mu} u_\mu = \rho u^\lambda.$$

This result can be interpreted by saying that the linear transformation defined by the momentum-energy tensor has the unit velocity vector as an eigenvector, the corresponding eigenvalue being ρ.

Let us now find the momentum-energy tensor for the case in which the continuous medium is a perfect fluid. Using the notation of (111.3) $T^{\lambda\mu}$ has the components

$$T^{ik} = t^{ik} = -p\eta^{ik}, \qquad T^{i4} = T^{4i} = T^{44} = 0$$

at the point P_0 in the frame S_0. In arbitrary coordinates, since $T^{\lambda\mu}$ satisfies (117.1), these take the tensor form

$$T^{\lambda\mu} = -p(g^{\lambda\mu} - u^\lambda u^\mu).$$

Hence

$$P^{\lambda\mu} = \left(\rho + \frac{p}{c^2}\right) u^\lambda u^\mu - \frac{p}{c^2} g^{\lambda\mu}. \tag{118.3}$$

V. THE MAXWELL-LORENTZ EQUATIONS

119. The electromagnetic field tensor. Maxwell's theory of the electromagnetic field in a vacuum and the Maxwell-Lorentz electron theory involve an electric field vector **E** and a magnetic field vector **H** which both vary with time. This representation is only useful, however, when considering transformations which consist of spatial rotations and translations and changes in the time origin. We shall now show how the Maxwell-Lorentz equations which govern the electromagnetic field may be expressed in a particularly simple way by using tensors in V_4. An important feature of this tensor representation is that it has led to a better understanding of the electrodynamics of moving bodies.

Consider a particular reduced Galilean coordinate system (x^1, x^2, x^3, x^4) which corresponds to a rectangular coordinate system $Oxyz$ in space. It is well known that the electric and magnetic fields can be defined in such a system by means of the vector potential $\mathbf{A}$ and the scalar potential Φ given by

$$\left. \begin{aligned} \mathbf{H} &= \operatorname{curl} \mathbf{A}, \\ \mathbf{E} &= -\operatorname{grad} \Phi - \frac{1}{c} \frac{\partial \mathbf{A}}{\partial t}. \end{aligned} \right\} \tag{119.1}$$

Let A_x, A_y, A_z represent the components of $\mathbf{A}$ along the axes $Oxyz$ and consider the *four*-vector $\boldsymbol{\varphi}$ which has the contravariant components

$$\phi^1 = -A_x, \qquad \phi^2 = -A_y, \qquad \phi^3 = -A_z, \qquad \phi^4 = -\Phi \tag{119.2}$$

in the coordinate system (x^1, x^2, x^3, x^4) of V_4. It therefore has the covariant components

$$\phi_1 = A_x, \quad \phi_2 = A_y, \quad \phi_3 = A_z, \quad \phi_4 = -\Phi. \tag{119.3}$$

The relations (119.1) can then be written in the form

$$\left. \begin{aligned} H_x &= \partial_2 \phi_3 - \partial_3 \phi_2, \\ H_y &= \partial_3 \phi_1 - \partial_1 \phi_3, \\ H_z &= \partial_1 \phi_2 - \partial_2 \phi_1, \end{aligned} \right\} \qquad \left. \begin{aligned} E_x &= \partial_1 \phi_4 - \partial_4 \phi_1, \\ E_y &= \partial_2 \phi_4 - \partial_4 \phi_2, \\ E_z &= \partial_3 \phi_4 - \partial_4 \phi_3. \end{aligned} \right\}$$

In other words, the six components of the vectors $\mathbf{H}$ and $\mathbf{E}$ in space are given by the six covariant components of the antisymmetric tensor

$$F_{\lambda\mu} = \partial_\lambda \phi_\mu - \partial_\mu \phi_\lambda \tag{119.4}$$

in V_4. In reduced Galilean coordinates we have explicitly

$$\left. \begin{aligned} H_x &= F_{23}, \\ H_y &= F_{31}, \\ H_z &= F_{12}, \end{aligned} \right\} \qquad \left. \begin{aligned} E_x &= F_{14}, \\ E_y &= F_{24}, \\ E_z &= F_{34}. \end{aligned} \right\} \tag{119.5}$$

The components of **H** and **E** can also be expressed using the contravariant components of the tensor $F_{\lambda\mu}$ as

$$
\left.\begin{array}{l}
H_x = F^{23}, \\
H_y = F^{31}, \\
H_z = F^{12},
\end{array}\right\}
\qquad
\left.\begin{array}{l}
E_x = -F^{14}, \\
E_y = -F^{24}, \\
E_z = -F^{34}.
\end{array}\right\}
\qquad (119.6)
$$

Thus in space-time the electromagnetic field is defined by an antisymmetric second-order tensor known as the electromagnetic field tensor. $F_{\lambda\mu}$ is the curl of ϕ_λ, which is called the *four-vector potential*. The introduction of the electromagnetic field tensor defined by (119.4) illustrates the fundamental unity of the electromagnetic field and removes the apparent independence of the electric and magnetic fields since, according to the tensor transformation law, the components of the electric field vector referred to one Galilean frame depend on both the electric *and* magnetic field components in any other frame.

Consider, for example, two systems of Galilean coordinates $(xyzt)$ and $(x'y'z't')$. It follows from the equations (108.12) defining the Lorentz transformation and from (119.5) that the components of the electric and magnetic fields in the second system are given, in terms of the components in the first system, by

$$
\begin{array}{ll}
H_{x'} = H_x, & E_{x'} = E_x, \\[2mm]
H_{y'} = \dfrac{H_y + \beta E_z}{\sqrt{(1-\beta^2)}}, & E_{y'} = \dfrac{E_y - \beta H_z}{\sqrt{(1-\beta^2)}}, \\[4mm]
H_{z'} = \dfrac{H_z - \beta E_y}{\sqrt{(1-\beta^2)}}, & E_{z'} = \dfrac{E_z + \beta H_y}{\sqrt{(1-\beta^2)}}.
\end{array}
$$

The tensor form, (119.4), of equations (119.1) permits a natural definition of the electromagnetic field in any system of space-time coordinates whether rectilinear or curvilinear.

It will be seen that the four-vector potential ϕ_λ is not uniquely determined by the electromagnetic field. The gradient of an arbitrary scalar field can be added to any four-vector potential satisfying (119.4) without modifying these equations. Thus ϕ_λ may be replaced by

$$
\phi_\lambda^* = \phi_\lambda + \partial_\lambda \Psi. \qquad (119.7)
$$

This procedure was called a *gauge transformation* by Hermann Weyl. Quantities, such as the electromagnetic field, which are left unchanged by a gauge transformation are said to be *gauge invariant*.

120. The adjoint electromagnetic field tensor. We saw in §52 that an arbitrary antisymmetric tensor can be uniquely associated with another antisymmetric tensor called its *adjoint*. The adjoint electromagnetic field tensor is of order $4 - 2 = 2$, and is defined by

$$F'^{\nu\rho} = \tfrac{1}{2}\epsilon^{\lambda\mu\nu\rho} F_{\lambda\mu}, \tag{120.1}$$

in any coordinate system.

Let space-time be referred to reduced Galilean coordinates. Then $g = -1$ and

$$\epsilon^{\lambda\mu\nu\rho} = \varepsilon^{\lambda\mu\nu\rho},$$

where $\varepsilon^{\lambda\mu\nu\rho}$ denotes the usual permutation indicator. In this way we get a simple physical interpretation of the components of the adjoint tensor in Galilean coordinates

$$\left.\begin{aligned} F'^{14} &= F_{23} = H_x, \\ F'^{24} &= F_{31} = H_y, \\ F'^{34} &= F_{12} = H_z, \end{aligned}\right\} \qquad \left.\begin{aligned} F'^{23} &= F_{14} = E_x, \\ F'^{31} &= F_{24} = E_y, \\ F'^{12} &= F_{34} = E_z. \end{aligned}\right\} \tag{120.2}$$

The adjoint tensor $F'^{\lambda\mu}$ is therefore an alternative representation of the electromagnetic field.

121. The electric current vector. The Maxwell-Lorentz theory of electrons, expressed in terms of a Galilean frame, is formulated in terms of the charge density ρ and the product of ρ with the velocity of the charge $\mathbf{v}$, i.e. $\rho\mathbf{v}$. Suppose the charge to be at rest with respect to a Galilean frame S_0 and let ρ_0 be the corresponding charge density which obviously defines a scalar in V_4.

The quantity of electricity de occupying the volume element dV_0 is given by

$$de = \rho_0 \, dV_0.$$

This quantity of electricity must be invariant in V_4. The total charge on an electron, for instance, is the same with respect to any Galilean frame. Thus, for an arbitrary Galilean frame S, we have

$$de = \rho \, dV,$$

where dV denotes the volume element in S which corresponds to dV_0 in S_0. It follows immediately from the equations of the Lorentz transformation (108.11) that, if S_0 is moving with respect to S with velocity $\mathbf{v}$, we have

$$dV = \sqrt{(1 - \beta^2)} \, dV_0.$$

Hence

$$\rho \sqrt{(1 - \beta^2)} \, dV_0 = \rho_0 \, dV_0,$$

and ρ is related to ρ_0 by

$$\rho = \frac{\rho_0}{\sqrt{(1 - \beta^2)}}. \tag{121.1}$$

The four-vector J^λ defined by the relation

$$J^\lambda = \rho_0 c u^\lambda \quad \left(u^\lambda = \frac{dy^\lambda}{ds} \right), \tag{121.2}$$

is called the *electric current vector*. Substituting the values (112.4) for u^λ, we find that

$$J^i = \frac{\rho_0 v^i}{\sqrt{(1 - \beta^2)}}, \qquad J^4 = \frac{\rho_0 c}{\sqrt{(1 - \beta^2)}},$$

hence, using (121.1),

$$J^i = \rho v^i, \qquad J^4 = \rho c. \tag{121.3}$$

122. The first set of Maxwell-Lorentz equations. The Maxwell-Lorentz equations can be split into two sets. The first of these is given in ordinary three-vector notation by

$$\operatorname{curl} \mathbf{H} - \frac{1}{c} \frac{\partial \mathbf{E}}{\partial t} = 4\pi \rho \frac{\mathbf{v}}{c}, \tag{122.1}$$

$$\operatorname{div} \mathbf{E} = 4\pi \rho. \tag{122.2}$$

The x component of (122.1) is given by

$$\frac{\partial H_z}{\partial y} - \frac{\partial H_y}{dz} - \frac{1}{c}\frac{\partial E_x}{\partial t} = 4\pi\frac{\rho v^1}{c},$$

and using (119.6) and (121.3) this may be rewritten

$$\partial_2 F^{12} + \partial_3 F^{13} + \partial_4 F^{14} = \frac{4\pi}{c}J^1.$$

We can therefore express (122.1) in reduced Galilean coordinates as

$$\partial_\mu F^{i\mu} = \frac{4\pi}{c}J^i. \tag{122.3}$$

Equation (122.2) can be written explicitly as

$$\frac{\partial E_x}{\partial x} + \frac{\partial E_y}{\partial y} + \frac{\partial E_z}{\partial z} = 4\pi\rho,$$

or using (119.6) and (121.3), as

$$\partial_1 F^{41} + \partial_2 F^{42} + \partial_3 F^{43} = \frac{4\pi}{c}J^4.$$

It follows that the first set of Maxwell-Lorentz equations can, using reduced Galilean coordinates, be written in the very simple form

$$\partial_\mu F^{\lambda\mu} = \frac{4\pi}{c}J^\lambda. \tag{122.4}$$

Since the coordinate system is rectilinear this is equivalent to

$$\nabla_\mu F^{\lambda\mu} = \frac{4\pi}{c}J^\lambda. \tag{122.5}$$

The tensor form (122.5) of Maxwell's equations (122.1) and (122.2) is also valid for any system of curvilinear coordinates.

123. The second set of Maxwell-Lorentz equations. Using Galilean coordinates, the other set of Maxwell-Lorentz equations are given, in three-vector notation, by

$$\operatorname{curl}\mathbf{E} + \frac{1}{c}\frac{\partial \mathbf{H}}{\partial t} = 0, \tag{123.1}$$

$$\operatorname{div} \mathbf{H} = 0. \tag{123.2}$$

The x-component of (123.1) gives

$$\frac{\partial E_z}{\partial y} - \frac{\partial E_y}{\partial z} + \frac{1}{c}\frac{\partial H_z}{\partial t} = 0,$$

or, using (120.2),

$$\partial_2 F'^{12} + \partial_3 F'^{13} + \partial_4 F'^{14} = 0.$$

Thus the vector relation (123.1) can be translated into reduced Galilean coordinates as

$$\partial_\mu F'^{i\mu} = 0. \tag{123.3}$$

Similarly (123.2) can be written

$$\frac{\partial H_x}{\partial x} + \frac{\partial H_y}{\partial y} + \frac{\partial H_z}{\partial z} = 0,$$

and, again using (120.2), this becomes

$$\partial_1 F'^{41} + \partial_2 F'^{42} + \partial_3 F'^{43} = 0.$$

It follows that the second set of Maxwell-Lorentz equations can be rewritten using reduced Galilean coordinates as

$$\partial_\mu F'^{\lambda\mu} = 0. \tag{123.4}$$

Since the coordinate system used is rectilinear this is equivalent to

$$\nabla_\mu F'^{\lambda\mu} = 0. \tag{123.5}$$

This tensor form is again valid for any system of coordinates. The similarity between (122.5) and (123.5) will be noted.

Equations (123.1) and (123.2) are simple consequences of (119.1); they merely express the fact that $\mathbf{E}$ and $\mathbf{H}$ are derivable from a vector potential $\mathbf{A}$ and a scalar potential Φ. Equations (123.5) can be written

$$\nabla_\mu(\epsilon^{\lambda\mu\nu\rho} F_{\nu\rho}) = 0. \tag{123.6}$$

However, remembering (69.3), we can easily verify that

$$\nabla_\mu \epsilon^{\lambda\mu\nu\rho} = 0.$$

Then equations (123.6) take the form

$$\epsilon^{\lambda\mu\nu\rho} \nabla_\mu F_{\nu\rho} = 0$$

and, on multiplying by $\sqrt{|g|}$, they become

$$\varepsilon^{\lambda\mu\nu\rho} \nabla_\mu F_{\nu\rho} = 0. \tag{123.7}$$

The covariant derivative $\nabla_\mu F_{\nu\rho}$ may be written

$$\nabla_\mu F_{\nu\rho} = \partial_\mu F_{\nu\rho} - \Gamma_\mu{}^\tau{}_\nu F_{\tau\rho} - \Gamma_\mu{}^\tau{}_\rho F_{\nu\tau}.$$

From considerations of symmetry

$$\varepsilon^{\lambda\mu\nu\rho} \Gamma_\mu{}^\tau{}_\nu = 0, \qquad \varepsilon^{\lambda\mu\nu\rho} \Gamma_\mu{}^\tau{}_\rho = 0,$$

hence, using an arbitrary system of coordinates, Maxwell's equations (123.5) may be written

$$\varepsilon^{\lambda\mu\nu\rho} \partial_\mu F_{\nu\rho} = 0. \tag{123.8}$$

These equations have the explicit form

$$\partial_\mu F_{\nu\rho} + \partial_\nu F_{\rho\mu} + \partial_\rho F_{\mu\nu} = 0,$$

where μ, ν, ρ denote three distinct indices. They are the necessary conditions for the existence of a four-vector ϕ_λ such that $F_{\lambda\mu}$ is its curl.

124. Charge conservation. Taking the divergence of (122.1), differentiating (122.2) and combining the equations thus obtained leads to the following expression for the law of charge conservation:

$$\frac{\partial \rho}{\partial t} + \operatorname{div}(\rho \mathbf{v}) = 0. \tag{124.1}$$

It is easy to reduce this law to its general tensor form. Consider first a system of reduced Galilean coordinates. The first set of Maxwell-Lorentz equations can, according to (122.4), be written as

$$\partial_\mu F^{\lambda\mu} = \frac{4\pi}{c} J^\lambda. \tag{124.2}$$

Differentiating with respect to x^λ (and carrying out the appropriate summation) we obtain

$$\partial_{\lambda\mu} F^{\lambda\mu} = \frac{4\pi}{c} \partial_\lambda J^\lambda.$$

However, taking account of the antisymmetry of $F^{\lambda\mu}$ and interchanging the repeated indices λ and μ, it follows that

$$\partial_{\lambda\mu}F^{\lambda\mu} = -\partial_{\lambda\mu}F^{\mu\lambda} = -\partial_{\mu\lambda}F^{\mu\lambda} = -\partial_{\lambda\mu}F^{\lambda\mu} = 0.$$

Hence, in reduced Galilean coordinates,

$$\partial_\lambda J^\lambda = 0, \tag{124.3}$$

which is an alternative form of (124.1). Referring space-time to an arbitrary curvilinear coordinate system, (124.3) becomes

$$\nabla_\lambda J^\lambda = 0. \tag{124.4}$$

This is the general tensor form of the law of charge conservation.

125. The Lorentz force density. Returning to Galilean coordinates, let **K** be the force density given by the Maxwell-Lorentz theory; that is the force exerted on the charge contained in unit volume. This is given in three-vector notation by the well-known formula

$$\mathbf{K} = \rho\mathbf{E} + \frac{\rho\mathbf{v}}{c} \wedge \mathbf{H}. \tag{125.1}$$

The x-component of (125.1) is

$$K_x = \rho E_x + \frac{\rho v_y}{c} H_z - \frac{\rho v_z}{c} H_y.$$

Expressing this relation in terms of the contravariant components of the electromagnetic field tensor and the covariant components of the current vector we get

$$K_x = \frac{1}{c}[J_4 F^{41} + J_2 F^{21} - J_3 F^{13}] = \frac{1}{c}J_\mu F^{\mu 1}.$$

The components K^i of the space vector **K** are therefore given by

$$K^i = \frac{1}{c}J_\mu F^{\mu i}.$$

A fourth component K^4 may be introduced to complete the set K^1, K^2, K^3. We then write

$$K^\lambda = \frac{1}{c}J_\mu F^{\mu\lambda}, \tag{125.2}$$

this tensor form being valid in an arbitrary coordinate system.

In terms of reduced Galilean coordinates the fourth component K^4 becomes

$$K^4 = \frac{1}{c}[J_1 F^{14} + J_2 F^{24} + J_3 F^{34}].$$

According to (125.1) this may be written in three-vector notation as

$$K^4 = \frac{1}{c}\rho\mathbf{v}\cdot\mathbf{E} = \frac{1}{c}\mathbf{K}\cdot\mathbf{v}.$$

Apart from the factor $1/c$, K^4 thus represents the work per unit time and volume which corresponds to the force density $\mathbf{K}$.

The four-vector defined by (125.2) is known as the *Lorentz force density*. Since $F^{\lambda\mu}$ is antisymmetric

$$K^\lambda J_\lambda = \frac{1}{c}F^{\lambda\mu}J_\lambda J_\mu = 0,$$

and we see that the force and current four-vectors are orthogonal.

126. The momentum-energy tensor of the electromagnetic field. Consider a continuous medium formed by the aggregate of a large number of charged particles – electrons for example. If the number is sufficiently large the effects due to individual particles will be negligible. We therefore consider the total electromagnetic field acting on an individual particle to be given independently of the behaviour of the particle and represent it by $F^{\lambda\mu}$. The continuous medium is thus subject to the Lorentz force density given by (125.2).

It follows that the equations of motion of the continuous medium are given by (118.2) with the substitution $\Phi^\lambda = K^\lambda$:

$$\nabla_\mu P^{\lambda\mu} = \frac{1}{c^2}K^\lambda. \tag{126.1}$$

It is possible to put the right-hand side of this equation into a similar form to that of the left-hand side. Substituting the current vector J^μ given by (122.5) into (125.2) we obtain

$$K_\lambda = -\frac{1}{4\pi}F_{\lambda\mu}\nabla_\rho F^{\mu\rho}.$$

Therefore

$$4\pi K_\lambda = -\nabla_\rho(F_{\lambda\mu}F^{\mu\rho}) + F^{\mu\rho}\nabla_\rho F_{\lambda\mu}.$$

The second term on the right-hand side gives

$$F^{\mu\rho}\nabla_\rho F_{\lambda\mu} = \tfrac{1}{2}(F^{\mu\rho}\nabla_\rho F_{\lambda\mu} + F^{\rho\mu}\nabla_\mu F_{\lambda\rho}) = \tfrac{1}{2}F^{\mu\rho}(\nabla_\rho F_{\lambda\mu} + \nabla_\mu F_{\rho\lambda}),$$

where we have interchanged repeated indices and used the anti-symmetry of $F_{\lambda\mu}$. However, from the second set of Maxwell's equations in the form (123.7), we have

$$\nabla_\rho F_{\lambda\mu} + \nabla_\mu F_{\rho\lambda} + \nabla_\lambda F_{\mu\rho} = 0$$

for any set of indices λ, μ, ρ. Hence

$$F^{\mu\rho}\nabla_\rho F_{\lambda\mu} = -\tfrac{1}{2}F^{\mu\rho}\nabla_\lambda F_{\mu\rho} = -\tfrac{1}{4}\nabla_\lambda(F^{\mu\rho}F_{\mu\rho}),$$

and we obtain the expression

$$-4\pi K_\lambda = \nabla_\rho(F_{\lambda\mu}F^{\mu\rho}) + \tfrac{1}{4}\nabla_\lambda(F_{\alpha\beta}F^{\alpha\beta})$$

for the Lorentz force density. Relabelling the indices we have

$$-4\pi K_\lambda = \nabla_\mu(-F_{\lambda\alpha}F^{\mu\alpha} + \tfrac{1}{4}g^\mu_\lambda F_{\alpha\beta}F^{\alpha\beta}).$$

We are thus led to introduce the symmetric tensor

$$M_{\lambda\mu} = \frac{1}{4\pi c^2}(\tfrac{1}{4}g_{\lambda\mu}F^{\alpha\beta}F_{\alpha\beta} - F_{\lambda\alpha}F_\mu{}^\alpha), \tag{126.2}$$

and equations (126.1) take the simple form

$$\nabla_\mu(P^{\lambda\mu} + M^{\lambda\mu}) = 0. \tag{126.3}$$

Since the right-hand side is zero, $M^{\lambda\mu}$ accounts for the electromagnetic interactions and may therefore be added to $P^{\lambda\mu}$ to obtain the total momentum-energy tensor of the continuous medium. $M_{\lambda\mu}$, defined by (126.2), is known as the *momentum-energy tensor of the electromagnetic field*. In reduced Galilean coordinates the components of this tensor comprise the components of the Maxwell stress tensor, the components of the Poynting vector and the energy density of the electromagnetic field.

CHAPTER VIII

Elements of the Relativistic Theory of Gravitation

127. Gravitation. The notion of gravitation is based essentially on the following experimental fact: all material bodies interact with one another at a distance. This interaction is described simply and precisely in terms of attractive forces within the framework of classical mechanics. Such a description is, however, not in accord with the relativistic requirements† previously introduced.

We shall refer to the law of motion of an infinitesimal test body, which is shielded from all electrical and contact interactions, as the *basic law of gravitation*. It is an experimental fact that, when the test body is not too far away from other material bodies, its motion differs appreciably from the uniform motion in a straight line which is implied by the principle of inertia. Consider the space-time manifold V_4 and suppose that it has the metric (111.1). We assume that the basic law of gravitation is directly linked with the form of ds^2 and is strictly determined by it. Then, as the Galilean ds^2 has constant coefficients, the basic law must be everywhere the same with respect to a Galilean frame. When the test body is far from other material bodies it satisfies the principle of inertia; this holds everywhere if the test body is completely isolated. Thus, according to the above hypothesis, a Galilean ds^2 is only capable of representing a universe devoid of matter, which is therefore also devoid of gravitation.

In order to represent a universe with gravitation, Einstein introduced metrics which were more general than the Minkowski metric (111.1) of special relativity theory. He considered *Riemannian space-times*, the metrics of which were supposed to determine the basic law of gravitation.

How is it possible to envisage such a relationship? We have seen in §112 that, in a universe without gravitation, the geodesics of the line

† For a discussion of this point see Chapter X of *Introduction to the Theory of Relativity*, by P. G. Bergmann (Prentice-Hall, 1942). (T.)

elements with positive ds^2 define the motion of a material test body and those for which $ds^2 = 0$ define the paths of light rays. We now extend this idea to the more general ds^2 introduced by Einstein. We postulate:

The geodesic principle: for any distribution of mass and energy, the geodesics of the line element of V_4 define the motions of material test bodies and the paths of light rays.

128. The metric of general relativity. In Einstein's conception, the universe is represented by a four-dimensional Riemannian space V_4 which has the metric

$$ds^2 = g_{\lambda\mu} \, dy^\lambda dy^\mu \qquad (128.1)$$

with signature $(+ \ - \ - \ -)$. ds is called the infinitesimal element of the interval in space-time. The physical interpretation of the coordinates and of the element of interval ds must be made in terms of the tangential or osculating Euclidean metrics at a point. At any particular point it is possible to bring such a Euclidean metric into the simple Minkowski form (111.3).

The ten $g_{\lambda\mu}$, which correspond to a particular system of coordinates, are functions of the y^λ. They define completely the basic law of gravitation in terms of this system by the geodesic principle. For this reason they are called the *gravitational potentials*. The derivatives of these potentials, which appear in the geodesic equations as the Christoffel symbols, define the gravitational field in the coordinate system under consideration.

The essential problem in the general theory of relativity is the determination of the gravitational potentials which correspond to various states of matter in motion.

129. Einstein's equations. Einstein was led to partial differential equations limiting the generality of the gravitational potentials by two essential requirements: on the one hand these equations must generalize the equations of Laplace and Poisson which govern the Newtonian potential; on the other hand they must be expressible in the form of relations between tensors in V_4. We assume that they may be written in the form of a relationship between two symmetric tensors

$$S_{\lambda\mu} = \chi Q_{\lambda\mu}, \tag{129.1}$$

where χ denotes a constant factor associated with a universal gravitational constant. The tensor $Q_{\lambda\mu}$ is taken to describe the energy distribution at each point (this being zero in matter free regions). It therefore generalizes the right-hand side of Poisson's equation.

In order to take account of all forms of energy we identify $Q_{\lambda\mu}$ with the total momentum-energy tensor which was studied in §**118** and §**126**. In the case of a continuous medium with electromagnetic interactions we therefore write

$$Q_{\lambda\mu} = P_{\lambda\mu} + M_{\lambda\mu}. \tag{129.2}$$

If we introduce an osculating Euclidean metric at a point in V_4 the results obtained in our study of special relativity become valid locally for the space-time of general relativity. In particular $Q_{\lambda\mu}$ satisfies equations (126.3) for the transfer of energy and momentum through a continuous medium, i.e.

$$\nabla_\mu Q^{\lambda\mu} = 0. \tag{129.3}$$

This equation may be interpreted by saying that the tensor $Q^{\lambda\mu}$ is *conservative*. It may be compared with (124.4) which represents the conservation of charge.

We assume the tensor $S^{\lambda\mu}$ appearing on the left-hand side of (129.1) to have a purely geometrical significance, which is to say that it only involves the metric (128.1) and its derivatives. *A priori* the ten $S_{\lambda\mu}$ cannot be independent. Disposing of the arbitrariness which exists in the coordinate system (y^λ) determines the values of four of the potentials. The remaining six potentials would then have to satisfy ten independent conditions if the quantities $S_{\lambda\mu}$ were completely independent.

Since $Q_{\lambda\mu}$ is conservative (129.1) implies that

$$\nabla_\mu S^{\lambda\mu} = 0, \tag{129.4}$$

which gives the four relations linking the ten functions $S_{\lambda\mu}$. We also expect the field equations to be generalizations of the second-order differential equations of Laplace and Poisson. In fact, the following two conditions are sufficient to determine the geometrical tensor $S_{\lambda\mu}$:

(*1*) The quantities $S_{\lambda\mu}$ depend only upon the gravitational potentials and their first and second-order derivatives. They are linear with respect to the second-order derivatives.

(2) The quantities $S_{\lambda\mu}$ satisfy the conservation equations

$$\nabla_{\mu} S^{\lambda\mu} = 0.$$

From the considerations of §85 and §86 we are already aware of a tensor which satisfies the above conditions. This is given by

$$R_{\lambda\mu} - \tfrac{1}{2} g_{\lambda\mu} R,$$

where $R_{\lambda\mu}$ is the Ricci tensor and R is the scalar Riemannian curvature of the Riemannian space (128.1). Élie Cartan has shown that the only tensors satisfying the above conditions are given by

$$S_{\lambda\mu} = h[R_{\lambda\mu} - \tfrac{1}{2} g_{\lambda\mu}(R+k)],$$

where h and k are constants. The corresponding partial differential equations may be written

$$R_{\lambda} - \tfrac{1}{2} g_{\lambda\mu}(R+k) = \chi Q_{\lambda\mu}$$

if we suppress the superfluous factor h. Except in certain very special cosmological studies one is only concerned in general relativity with equations where k is zero (which were first given by Einstein). Einstein's gravitational equations are therefore given by

$$R_{\lambda\mu} - \tfrac{1}{2} g_{\lambda\mu} R = \chi Q_{\lambda\mu} \tag{129.5}$$

or, in the absence of gravitating matter,

$$R_{\lambda\mu} - \tfrac{1}{2} g_{\lambda\mu} R = 0. \tag{129.6}$$

Equations (129.6) are equivalent to

$$R_{\lambda\mu} = 0, \tag{129.7}$$

as these equations immediately imply that $R = 0$.

130. The momentum-energy tensor. The momentum-energy tensor $Q_{\lambda\mu}$ which appears on the right-hand side of the Einstein equations should describe completely the distribution of matter and energy in motion. It is necessary, as in Newtonian mechanics, to be content with an incomplete description which corresponds to a tensor $Q_{\lambda\mu}$ of restricted complexity. This tensor contains various terms which correspond to the different kinds of energy: rest-mass energy, kinetic energy, energy deriving from stresses, and from the electromagnetic field etc. In the presence of matter the term which is experimentally the most important is that which corresponds to the rest-mass energy.

We note that, because of the non-uniformity of Riemannian space, one cannot suppose material bodies to be rigid as in classical mechanics. We are thus forced to consider models which are hydro-dynamical in origin; this is why the study of continuous media plays such an important role in the relativistic theory of gravitation.

Many forms of the momentum-energy tensor are commonly used in relativistic gravitational theory. Consider first a continuous medium such that interactions in the interior of the medium are negligible and there are no electromagnetic interactions. The corresponding form of the momentum-energy tensor is given by

$$Q^{\lambda\mu} = \rho u^\lambda u^\mu, \qquad (130.1)$$

where ρ is the density of the matter at rest and where u^λ is the unit velocity vector of the matter distribution at the point considered. The tensor (130.1) corresponds only to the rest mass and kinetic energy.

If interactions defined by a stress tensor $T_{\lambda\mu}$ exist inside the medium we have

$$Q^{\lambda\mu} = \rho u^\lambda u^\mu + \frac{1}{c^2} T^{\lambda\mu}. \qquad (130.2)$$

In particular, in the case of a perfect fluid (118.3) gives

$$Q^{\lambda\mu} = \left(\rho + \frac{p}{c^2}\right) u^\lambda u^\mu - \frac{p}{c^2} g^{\lambda\mu}, \qquad (130.3)$$

where p denotes the pressure.

The lines in V_4 which are everywhere tangential to the unit velocity vector u^λ are called the *flow lines* of the distribution of matter.

The electromagnetic field in the absence of matter gives rise to the momentum-energy tensor defined by (126.2).

131. Conservation equations inside matter. Let us apply the conservation equations

$$\nabla_\mu Q^{\lambda\mu} = 0 \tag{131.1}$$

to the momentum-energy tensor given by (130.1): it follows that

$$\nabla_\mu(\rho u^\mu)\, u^\lambda + \rho u^\mu \nabla_\mu u^\lambda = 0. \tag{131.2}$$

However, since u^λ is a unit vector, we have

$$u_\lambda \nabla_\mu u^\lambda = 0.$$

Hence, after contracted multiplication by u_λ, (131.2) gives

$$\nabla_\mu(\rho u^\mu) = 0, \tag{131.3}$$

and

$$u^\mu \nabla_\mu u^\lambda = 0. \tag{131.4}$$

The geometrical interpretation of these equations is very simple: (131.3) states that the divergence of the vector ρu^μ, the generalized matter flux, is zero; this is the equation of continuity for the matter. (131.4) states that the flow lines are geodesics of ds^2, this result being directly related to the geodesic principle. Analogous but more complicated results have been given by Eisenhart and Lichnerowicz for the case in which there is an internal stress in the medium.

In view of the scope of this work we are forced to restrict the present discussion to these relatively elementary considerations, and to refer the reader who desires a more detailed account to the specialized treatises listed in the bibliography.

Bibliography

TENSOR CALCULUS

(1) *Lectures on Linear Algebra:* I. M. Gel'fand (Interscience; 1961)
(2) *Tensor Calculus:* B. Spain (Oliver and Boyd; 1953)
(3) *Vector and Tensor Analysis:* L. Brand (Wiley; 1947)
(4) *Riemannian Geometry:* L. P. Eisenhart (Princeton; 1926)
(5) *The Absolute Differential Calculus:* T. Levi-Civita (Blackie; 1926)
(6) *Tensor Calculus:* J. L. Synge and A. Schild (Univ. of Toronto Press; 1956)

Remarks: (1) provides a detailed treatment of the concept of a vector space and concludes with a brief introduction to tensor theory. (2) is a concise elementary textbook on tensor calculus and its applications which emphasises the manipulative aspects of the subject. (3), (4), (5) and (6) are standard treatises on tensor calculus. (3) includes several unusual topics; (5) contains a good introduction to relativity theory.

GENERAL APPLICATIONS

(1) *Les tenseurs en méchanique et en élasticité:* L. Brillouin (Masson; 1938)
(2) *Tensors for Circuits:* G. Kron (Dover; 1959)
(3) *Theoretical Elasticity:* A. E. Green and W. Zerna (Oxford; 1954)
(4) *Mathematical Theory of Elasticity:* I. S. Sokolnikoff (McGraw Hill; 1946)

Remarks: There are apparently no books on the use of tensors in analytical mechanics available in English. (1) is a treatise on this subject and the use of tensors in elasticity theory. (2) is primarily meant for electrical engineers. It will, however, give the reader an insight into one of the more exotic applications of tensor calculus. (3) and (4) are standard treatises.

RELATIVITY THEORY

(1) *Special Relativity:* W. Rindler (Oliver and Boyd; 1960)
(2) *Special Theory of Relativity:* J. Aharoni (Oxford U.P.; 1959)

160

(3) *Relativity; the Special Theory:* J. L. Synge (North Holland; 1956)

(4) *Relativity; the General Theory:* J. L. Synge (North Holland; 1960)

(5) *General Relativity and Gravitational Waves:* J. Weber (Interscience; 1961)

(6) *Théories relativistes de la gravitation et de l'électro-magnétisme:* A. Lichnerowicz (Masson; 1954)

(7) *Relativity, Thermodynamics and Cosmology:* R. C. Tolman (Oxford U.P.; 1950)

(8) *The Meaning of Relativity:* A. Einstein (Methuen; 1950)

(9) *The Theory of Space, Time and Gravitation:* V. Fock (Pergamon Press; 1958)

(10) *Theory of Relativity:* W. Pauli (Pergamon Press; 1958)

(11) *Introduction to the Theory of Relativity:* P. G. Bergmann (Prentice Hall; 1942)

(12) *The Theory of Relativity:* C. Møller (Oxford U.P.; 1952)

Remarks: (1) is an elementary textbook on special relativity theory. (2) is an algebraic treatise which provides a thorough grounding in the relativity theory used in quantum field theories. (3) and (4) are modern treatises written (in a most stimulating and colourful style) from a geometrical point of view. (5) sets out concisely the main aspects of modern research in general relativity theory. The remaining books, (6) to (12), are standard treatises with an emphasis on general relativity.

Index

Mathematics-Bestsellers

HANDBOOK OF MATHEMATICAL FUNCTIONS: with Formulas, Graphs, and Mathematical Tables, Edited by Milton Abramowitz and Irene A. Stegun. A classic resource for working with special functions, standard trig, and exponential logarithmic definitions and extensions, it features 29 sets of tables, some to as high as 20 places. 1046pp. 8 x 10 1/2. 0-486-61272-4

ABSTRACT AND CONCRETE CATEGORIES: The Joy of Cats, Jiri Adamek, Horst Herrlich, and George E. Strecker. This up-to-date introductory treatment employs category theory to explore the theory of structures. Its unique approach stresses concrete categories and presents a systematic view of factorization structures. Numerous examples. 1990 edition, updated 2004. 528pp. 6 1/8 x 9 1/4. 0-486-46934-4

MATHEMATICS: Its Content, Methods and Meaning, A. D. Aleksandrov, A. N. Kolmogorov, and M. A. Lavrent'ev. Major survey offers comprehensive, coherent discussions of analytic geometry, algebra, differential equations, calculus of variations, functions of a complex variable, prime numbers, linear and non-Euclidean geometry, topology, functional analysis, more. 1963 edition. 1120pp. 5 3/8 x 8 1/2. 0-486-40916-3

INTRODUCTION TO VECTORS AND TENSORS: Second Edition--Two Volumes Bound as One, Ray M. Bowen and C.-C. Wang. Convenient single-volume compilation of two texts offers both introduction and in-depth survey. Geared toward engineering and science students rather than mathematicians, it focuses on physics and engineering applications. 1976 edition. 560pp. 6 1/2 x 9 1/4. 0-486-46914-X

AN INTRODUCTION TO ORTHOGONAL POLYNOMIALS, Theodore S. Chihara. Concise introduction covers general elementary theory, including the representation theorem and distribution functions, continued fractions and chain sequences, the recurrence formula, special functions, and some specific systems. 1978 edition. 272pp. 5 3/8 x 8 1/2. 0-486-47929-3

ADVANCED MATHEMATICS FOR ENGINEERS AND SCIENTISTS, Paul DuChateau. This primary text and supplemental reference focuses on linear algebra, calculus, and ordinary differential equations. Additional topics include partial differential equations and approximation methods. Includes solved problems. 1992 edition. 400pp. 7 1/2 x 9 1/4. 0-486-47930-7

PARTIAL DIFFERENTIAL EQUATIONS FOR SCIENTISTS AND ENGINEERS, Stanley J. Farlow. Practical text shows how to formulate and solve partial differential equations. Coverage of diffusion-type problems, hyperbolic-type problems, elliptic-type problems, numerical and approximate methods. Solution guide available upon request. 1982 edition. 414pp. 6 1/8 x 9 1/4. 0-486-67620-X

VARIATIONAL PRINCIPLES AND FREE-BOUNDARY PROBLEMS, Avner Friedman. Advanced graduate-level text examines variational methods in partial differential equations and illustrates their applications to free-boundary problems. Features detailed statements of standard theory of elliptic and parabolic operators. 1982 edition. 720pp. 6 1/8 x 9 1/4. 0-486-47853-X

LINEAR ANALYSIS AND REPRESENTATION THEORY, Steven A. Gaal. Unified treatment covers topics from the theory of operators and operator algebras on Hilbert spaces; integration and representation theory for topological groups; and the theory of Lie algebras, Lie groups, and transform groups. 1973 edition. 704pp. 6 1/8 x 9 1/4. 0-486-47851-3

Browse over 9,000 books at www.doverpublications.com

A SURVEY OF INDUSTRIAL MATHEMATICS, Charles R. MacCluer. Students learn how to solve problems they'll encounter in their professional lives with this concise single-volume treatment. It employs MATLAB and other strategies to explore typical industrial problems. 2000 edition. 384pp. 5 3/8 x 8 1/2.   0-486-47702-9

NUMBER SYSTEMS AND THE FOUNDATIONS OF ANALYSIS, Elliott Mendelson. Geared toward undergraduate and beginning graduate students, this study explores natural numbers, integers, rational numbers, real numbers, and complex numbers. Numerous exercises and appendixes supplement the text. 1973 edition. 368pp. 5 3/8 x 8 1/2.   0-486-45792-3

A FIRST LOOK AT NUMERICAL FUNCTIONAL ANALYSIS, W. W. Sawyer. Text by renowned educator shows how problems in numerical analysis lead to concepts of functional analysis. Topics include Banach and Hilbert spaces, contraction mappings, convergence, differentiation and integration, and Euclidean space. 1978 edition. 208pp. 5 3/8 x 8 1/2.   0-486-47882-3

FRACTALS, CHAOS, POWER LAWS: Minutes from an Infinite Paradise, Manfred Schroeder. A fascinating exploration of the connections between chaos theory, physics, biology, and mathematics, this book abounds in award-winning computer graphics, optical illusions, and games that clarify memorable insights into self-similarity. 1992 edition. 448pp. 6 1/8 x 9 1/4.   0-486-47204-3

SET THEORY AND THE CONTINUUM PROBLEM, Raymond M. Smullyan and Melvin Fitting. A lucid, elegant, and complete survey of set theory, this three-part treatment explores axiomatic set theory, the consistency of the continuum hypothesis, and forcing and independence results. 1996 edition. 336pp. 6 x 9.   0-486-47484-4

DYNAMICAL SYSTEMS, Shlomo Sternberg. A pioneer in the field of dynamical systems discusses one-dimensional dynamics, differential equations, random walks, iterated function systems, symbolic dynamics, and Markov chains. Supplementary materials include PowerPoint slides and MATLAB exercises. 2010 edition. 272pp. 6 1/8 x 9 1/4.   0-486-47705-3

ORDINARY DIFFERENTIAL EQUATIONS, Morris Tenenbaum and Harry Pollard. Skillfully organized introductory text examines origin of differential equations, then defines basic terms and outlines general solution of a differential equation. Explores integrating factors; dilution and accretion problems; Laplace Transforms; Newton's Interpolation Formulas, more. 818pp. 5 3/8 x 8 1/2.   0-486-64940-7

MATROID THEORY, D. J. A. Welsh. Text by a noted expert describes standard examples and investigation results, using elementary proofs to develop basic matroid properties before advancing to a more sophisticated treatment. Includes numerous exercises. 1976 edition. 448pp. 5 3/8 x 8 1/2.   0-486-47439-9

THE CONCEPT OF A RIEMANN SURFACE, Hermann Weyl. This classic on the general history of functions combines function theory and geometry, forming the basis of the modern approach to analysis, geometry, and topology. 1955 edition. 208pp. 5 3/8 x 8 1/2.   0-486-47004-0

THE LAPLACE TRANSFORM, David Vernon Widder. This volume focuses on the Laplace and Stieltjes transforms, offering a highly theoretical treatment. Topics include fundamental formulas, the moment problem, monotonic functions, and Tauberian theorems. 1941 edition. 416pp. 5 3/8 x 8 1/2.   0-486-47755-X

Browse over 9,000 books at www.doverpublications.com

Mathematics–Logic and Problem Solving

PERPLEXING PUZZLES AND TANTALIZING TEASERS, Martin Gardner. Ninety-three riddles, mazes, illusions, tricky questions, word and picture puzzles, and other challenges offer hours of entertainment for youngsters. Filled with rib-tickling drawings. Solutions. 224pp. 5 3/8 x 8 1/2. 0-486-25637-5

MY BEST MATHEMATICAL AND LOGIC PUZZLES, Martin Gardner. The noted expert selects 70 of his favorite "short" puzzles. Includes The Returning Explorer, The Mutilated Chessboard, Scrambled Box Tops, and dozens more. Complete solutions included. 96pp. 5 3/8 x 8 1/2. 0-486-28152-3

THE LADY OR THE TIGER?: and Other Logic Puzzles, Raymond M. Smullyan. Created by a renowned puzzle master, these whimsically themed challenges involve paradoxes about probability, time, and change; metapuzzles; and self-referentiality. Nineteen chapters advance in difficulty from relatively simple to highly complex. 1982 edition. 240pp. 5 3/8 x 8 1/2. 0-486-47027-X

SATAN, CANTOR AND INFINITY: Mind-Boggling Puzzles, Raymond M. Smullyan. A renowned mathematician tells stories of knights and knaves in an entertaining look at the logical precepts behind infinity, probability, time, and change. Requires a strong background in mathematics. Complete solutions. 288pp. 5 3/8 x 8 1/2.
0-486-47036-9

THE RED BOOK OF MATHEMATICAL PROBLEMS, Kenneth S. Williams and Kenneth Hardy. Handy compilation of 100 practice problems, hints and solutions indispensable for students preparing for the William Lowell Putnam and other mathematical competitions. Preface to the First Edition. Sources. 1988 edition. 192pp. 5 3/8 x 8 1/2. 0-486-69415-1

KING ARTHUR IN SEARCH OF HIS DOG AND OTHER CURIOUS PUZZLES, Raymond M. Smullyan. This fanciful, original collection for readers of all ages features arithmetic puzzles, logic problems related to crime detection, and logic and arithmetic puzzles involving King Arthur and his Dogs of the Round Table. 160pp. 5 3/8 x 8 1/2.
0-486-47435-6

UNDECIDABLE THEORIES: Studies in Logic and the Foundation of Mathematics, Alfred Tarski in collaboration with Andrzej Mostowski and Raphael M. Robinson. This well-known book by the famed logician consists of three treatises: "A General Method in Proofs of Undecidability," "Undecidability and Essential Undecidability in Mathematics," and "Undecidability of the Elementary Theory of Groups." 1953 edition. 112pp. 5 3/8 x 8 1/2. 0-486-47703-7

LOGIC FOR MATHEMATICIANS, J. Barkley Rosser. Examination of essential topics and theorems assumes no background in logic. "Undoubtedly a major addition to the literature of mathematical logic." – *Bulletin of the American Mathematical Society.* 1978 edition. 592pp. 6 1/8 x 9 1/4. 0-486-46898-4

INTRODUCTION TO PROOF IN ABSTRACT MATHEMATICS, Andrew Wohlgemuth. This undergraduate text teaches students what constitutes an acceptable proof, and it develops their ability to do proofs of routine problems as well as those requiring creative insights. 1990 edition. 384pp. 6 1/2 x 9 1/4. 0-486-47854-8

FIRST COURSE IN MATHEMATICAL LOGIC, Patrick Suppes and Shirley Hill. Rigorous introduction is simple enough in presentation and context for wide range of students. Symbolizing sentences; logical inference; truth and validity; truth tables; terms, predicates, universal quantifiers; universal specification and laws of identity; more. 288pp. 5 3/8 x 8 1/2. 0-486-42259-3

Browse over 9,000 books at www.doverpublications.com